English Nature Science

Number 34

Long-term studies in British woodland

K.J. Kirby[1]
M.D. Morecroft[2]

[1] English Nature
Northminster House
Peterborough
PE1 1UA

[2] Environmental Change Network
Institute of Terrestrial Ecology
Oxford University Field Laboratory
Wytham
Oxford
OX2 8QT

ISBN 1 85716 507 1

We would like to dedicate this volume to two former members of the Oxford Forestry Department, Colyear Dawkins and Eustace Jones and their pioneering works at Wytham and Lady Park Wood respectively.

PREFACE

The papers in this volume were presented at the British Ecological Society symposium held in the Oxford Forestry Institute on 13 July 1999. Our aim was to bring together examples of different types of long-term monitoring. The papers and posters presented explore changes in British woods over periods ranging from five to 50 years in trees and shrubs, ground flora, invertebrates and birds. The work described deals both with natural processes such as stand dynamics, population cycles, phenology and interactions of plant host and invertebrates; and with human-induced effects - woodland management, air pollution and climate change.

As important as the results of the studies themselves are the common threads that emerge with respect to the establishment and maintenance of long-term studies; the archiving and publishing of results so that others may continue the work; and the tension between trying to design systems that might be useful to future researchers, but are also simple enough that they do survive to be picked up by those (hypothetical) researchers.

We hope that these papers will provide inspiration and encouragement for those who are struggling with potential long-term studies of their own.

ACKNOWLEDGEMENTS

The meeting was organised under the auspices of the British Ecological Society's Forest Ecology Group with support from English Nature and the Environmental Change Network. The Oxford Forestry Institute kindly hosted the meeting. Special thanks go to Mary Roberts who has dealt with the preparation of the final versions of the papers.

Keith Kirby, English Nature
Mike Morecroft, NERC Institute of Terrestrial Ecology

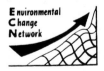

Recent English Nature and JNCC publications relevant to British woodland studies

English Nature Research Reports

43. PARKER, S.J. & WHITBREAD, A.M. 1993. Re-recording storm damaged woods in Kent and Sussex.

143. KIRBY, K.J., THOMAS, R.C. & DAWKINS, H.C. 1995. Changes in the structure and composition of the tree and shrub layers in Wytham Woods (Oxfordshire), 1974-1991.

175. KIRBY, K.J. & BELL, J. 1996. Changes in abundance of six ground flora species in Wytham Woods (1974-1991).

270. MOUNTFORD, E.P. & PETERKEN, G.F. 1998. Monitoring natural stand change in Monks Wood National Nature Reserve.

301. SOLLY, L., KIRBY, K.J. & SODEN, N, D. 1999. National sample survey of SSSI woodland.

302. MOUNTFORD, E.P., PETERKEN, G.F. & BURTON, D. 1998. Long-term monitoring and management of Langley Wood.

320. KIRBY, K.J. & THOMAS, R.C. 1999. Changes in the ground flora in Wytham Woods, southern England, 1974-1991, and their implications for nature conservation.

324. KIRBY, K.J. 1999. Woodland surveillance and monitoring - a discussion paper.

340. HOPKINS, B. 1999. The effect of shade and weather on daffodils *Narcissus pseudonarcissus* in West Dean Woods, West Sussex.

346. MOUNTFORD, E.P. & PETERKEN, G.F. 2000. Natural developments at Scords Wood, Toy's Hill, Kent since the Great Storm of October 1987.

348. MOUNTFORD, E.P., PAGE, P.A. & PETERKEN, G.F. 2000. Twenty-five years of change in a population of oak saplings in Wistman's Wood, Devon.

English Nature Science

23. KIRBY, K.J. & BUCKLEY, G.P. 1994. Ecological responses to the 1987 Great Storm in the woods of south-east England.

34. KIRBY, K.J. & MORECROFT, K.D. 2000. Long term studies in British woodland (this volume).

JNCC Reports

295. HALL, J.E., KIRBY, K.J. & MORECROFT, M.D. 1999. Minimum intervention woodlands and their use for ecological research in Great Britain.

298. PURDY, K.M. & FERRIS, R. 1999. A pilot study to examine the potential linkage between and applications of multiple woodland datasets: a GIS based approach.

300. BUNCE, R.G.H. & HIRST, N. 2000. Woodland change 1971-1988 in north-west England and south-west Scotland - a trial survey.

Contents

THE HUMAN ELEMENT IN LONG TERM WOODLAND STUDIES

G.F. Peterken
Beechwood House, St Briavels Common, Lydney, Gloucestershire GL15 6SL

Summary

Experience with a number of long-term studies is summarized and key differences between these and short-term research projects are identified. If new studies are to survive to become long-term they must be designed around human limitations, with as much attention paid to archiving and institutional memory as to ecology and research design.

Introduction

Long-term ecological studies are not like ecological research that lasts, say, less than five years (Taylor 1989). Short-term research projects are presented as studies that started with an hypothesis, continued with observations or experiments carefully designed to falsify that hypothesis, reached a conclusion, then stopped. Long-term studies on the other hand last much longer, by definition, tend to proceed by observation and inference, and remain open-ended. Almost invariably, short-term studies remain closely under the control of the initiators, but long-term studies are inevitably subject to the vagaries of changing institutional and personal interests and priorities, and must eventually be relinquished by those who started them.

This paper is based on several studies of change in the composition and structure of unmanaged semi-natural woods in Britain. These include both studies we initiated in the 1980s, and others started decades before by other ecologists (Peterken & Backmeroff 1988). Their survival and achievements have been limited by human failings of various kinds, which are likely to remain a factor even in the modern era of recording and data storage. Recognition of the limitations, however, many help those who design long-term studies in the future.

Woodland processes *v* human time scales

It may be helpful to compare the rate and duration of woodland processes with human spans. Dominant trees, such as oak *Quercus* spp., beech *Fagus sylvatica* and lime *Tilia* spp., may live for 300-500 years. The intervals between major natural disturbances may be decades or centuries. For example, the great storm of 1987 in south-east England was said to be the most severe since 1703. When a managed mixed-deciduous wood is allowed to develop naturally, gap formation may start within decades, but dead wood volumes will probably take 100 years or so to build up to natural levels.

Admittedly, significant changes can take place quickly. For example, mature stands may be levelled by a storm within an hour or so. The colonisation of Monks Wood NNR by muntjac deer *Muntiacus reevesii* resulted in profound changes in stand structure and ground flora composition within a decade (Crampton *et al*. 1998). The 1976 drought in Lady Park Wood killed many beech within a year or two, and stopped the survivors from growing for perhaps 8 years (Peterken & Mountford 1996). Disease killed most adult elm trees *Ulmus glabra* within five years (Peterken & Mountford 1998). However, in all these instances, the consequences of the short-term event could only be appreciated against a background of long-term observations, including records of prior condition and subsequent responses, which may need to be made over periods of at least 20 years.

Contrast woodland processes with human time-spans. Research grants last for about 3 years, and funding for any project rarely lasts longer than 5 years. Ecologists tend to change their jobs every 5-10 years. Even if individuals stay in one job, their interests and priorities tend to change over a decade. A career in ecology lasts for 25-35 years. Even if an individual maintains an interest throughout a career, institutional priorities change, and institutions themselves are regularly re-organised. For example, my own 23-year career as a specialist woodland ecologist for the Government's nature conservation organisation involved working successively for the Nature Conservancy, Natural Environment Research Council, Institute of Terrestrial Ecology, Nature Conservancy Council, English Nature and the Joint Nature Conservation Committee, three changes of location and six changes of office (which is when old records are easily lost), all without once changing the character of my work.

The mis-match between human and woodland timescales extends beyond individuals and institutions. Ecology has only been recognised as a distinct science for about a century. Modern forestry dates back further, but it was rare - but not unknown - for early foresters to record observations that are valuable to ecologists today, eg in Boubinsky Prales (Peterken 1981). Moreover, intellectual fashion changes in decades, or quicker. Any woodland study that has endured for 50 years will have been maintained against a background of loss of confidence in the concept of succession to a stable climax, and the rise in consciousness of the importance of natural disturbances.

How long does it take before studies become 'long-term'?

It is rarely worth re-recording studies of stand change in less than 10 years. Change is usually fairly slow, and errors in tree measurement are large in relation to growth over a few years. However, by 10 years, for example, significant relationships between growth rates and size can be detected, for example at Langley Wood (Mountford *et al*. 1998) and Craigellachie birchwood (Mountford and Peterken, unpublished data). On this basis, a decade is required to measure initial rates and directions of change, and 20 years is needed for a third recording, which allows any changes in rates and direction to be measured.

Even in 20 years, one is 'lucky' to observe a disturbance. At Clairinsh a hurricane passed seven years after the first recording (Backmeroff & Peterken 1989), and in Denny Wood regeneration ceased about a decade after observations started (Mountford *et al*. 1999), but in Lady Park Wood the first substantial events, the elm disease outbreak of about 1971 and the drought of 1976, occurred 26 years and 31 years respectively after the first records (Peterken & Jones 1987). The cessation of regeneration at Denny only became obvious nearly 30 years after recording started. On the basis that over the last 50 years a majority of the woods that I have helped to study have sustained a major event, such as a blowdown, severe drought, disease outbreak or major change in grazing pressure, a long-term study has an even chance of observing a major event in, perhaps, 35-40 years.

A general threshold of 25 years seems appropriate. If a study endures for longer than that, it can reasonably be described as 'long-term'.

How do studies become long-term ?

If any woodland study is to become long-term it must survive at least one transition from initiator to successors. It must achieve this without guaranteed funding, against a background of changing concepts and institutions. It will only do this if:

- the record is understandable by a successor who will probably not have access to the initiator;

- the record itself survives and can be understood;

- someone knows the record survives;

- the study site itself survives in a state that is worth studying; and

- present-day ecologists think the study is worth maintaining and recording.

In other words, if someone is planning to establish a new long-term study, they have to pay attention to (i) recording clearly and completely the methods used, (ii) archiving, (iii) the institutional memory, (iv) site management, and (v) they will also have to record features which have enduring value for their successors.

These points may be obvious, but many potentially long-term studies have failed because one of these conditions was not fulfilled, and those that have survived have been vitiated by partial failures. The history of five studies that had endured for at least 25 years was told by Peterken and Backmeroff (1988), and since then the tale of human failings in these studies has continued, as exemplified by the study in Denny Wood (Mountford *et al.* 1999). Since 1988 the entire original record of the Denny transect (started 1956) was lost when someone moved jobs: fortunately in this instance the record had been copied to another institution. Moreover, it has lately become clear that a recording in the 1970s has been completely lost, and that this would have been extremely valuable for estimating the effect of the 1976 drought. On the other hand, a second transect, which had last been recorded in 1964, was relocated in 1999 after many years of searching. At Wistmans Wood a complete copy of the records (started 1922) was lost within an NCC office, and the archive now survives as a single copy kept in a box in my loft. In the Black Wood of Rannoch one of the five plots initiated in 1948 by the Forestry Commission was accidentally partially 'overprinted' by an exclosure paid for from a research allocation which I controlled. At Lady Park Wood, the value of the 55-year observations is being compromised by excessively high deer populations: what had been a study of near-natural woodland is becoming a study of conversion from high forest to wood-pasture.

One of the key features of records that have survived to be understood and useful is simplicity. Simple measurements of tree presence and size (girth) become valuable in the long-term in much the same way that one pound, invested at average rates of interest, would become extremely valuable within 100-200 years. Elaborate codified observations (eg of canopy stratification) become unintelligible if the code is lost, and useless if successors have no faith in the code. For our studies, we favour transects over random or grids of plots, even though separate plots may be statistically superior, partly because transects are more easily found and reconstructed by successors after a period of apathy, whereas lost plots tend to be lost for ever.

Perception of long-term studies

In addition to the accidents of recording, data storage, etc, two other factors have, I think, made long-term studies unpopular amongst professionals. The results are often perceived as boring, predictable, and answering the questions posed several decades ago, which reduces their scientific status. Even if they are regarded as valuable, patience is required, yet most ecologists want to move on to other interesting topics or into senior management.

The predictability criticism has some force. Thus, after 10 years, the second recordings in Craigellachie, Black Wood of Rannoch, Denny Wood, Glen Tanar, Lady Park Wood and Langley Wood all showed patterns of growth and thinning in undisturbed stands that were entirely predictable. If Wistmans Wood had been recorded again after 10 years, it would hardly have changed at all. Even where a significant change occurred within the first 10 years (eg the invasion of Monks Wood by muntjac), the permanent plot records merely quantified changes that were already obvious.

Nevertheless, with the passage of time and the completion of a third recording, it becomes possible to measure changes in the rates of processes, such as growth or mortality. The chance that a formative event will take place within the period of observation increases with time. Furthermore, as time passes, stands approach old-growth, and the chance that a particular event will have a significant impact will increase. Thus, for example, the incidence of disturbances within the old-growth stands at Lady Park Wood has increased over the 55 years of observation, not because the wood has lately been more closely observed, but because the stand is more mature and vulnerable to disturbance.

Table 1. Significant events in studies lasting more than 35 years, and the time that has elapsed between the start of the study and the time when the event or the impact became obvious

Wistmans Wood (started 1921)	Grazing change and period of regeneration, ended c1965 (44 years). Rowan de-barking during prolonged snow cover, 1962/3 (41 years). Heavy snow storm, 1977/8 (56 years). Increasingly vigorous growth, which may or may not reflect climatic change (c60 years).
Lady Park Wood (started 1945)	Instability of large trees on slope (2 weeks). Grey squirrel impacts (14 and 38 years). Dutch elm disease (25-30 years). Severe drought (32 years). Late spring snowfall (39 years). Catastrophic vole damage to regeneration (40 years). Impact of excessive fallow deer populations (45 years).
Black Wood of Rannoch (started 1948)	Storm damage (c20 years).
Denny Wood (started 1955)	Stress due to wet and dry summers (5 years). Effects of increased grazing/browsing on regeneration (10-20 years). Severe drought (22 years). Grey squirrel damage becomes severe (c25 years). Storm impacts (32 years).
Clairinsh (started 1961)	Hurricane impacts (7 years).

In practice, projects that have survived for 25 years or more have all sustained significant events, none of which were predictable at the outset (Table 1). Yet the impacts of the events have only become apparent many years later, eg the cessation of regeneration in Denny Wood about 1964. Even where the immediate impacts were very obvious, the response of the stand may still not be complete after 20 years (eg beeches in Lady Park Wood are still dying from the 1976 drought in 1999). On the other hand, the impact of the 1968 hurricane at Clairinsh has turned out to be small, though at the time it looked at least locally devastating.

The scientific value of long-term studies has been elaborated elsewhere (Strayer *et al.* 1986). Rare and unpredictable events can be studied, eg the arrival of elm disease in Lady Park Wood (Peterken & Mountford 1998). The significance of short-term phenomena can be seen, eg the four-hour late-April snow-lie that has permanently altered the form of limes in Lady Park Wood and would have increased their rate of vegetative reproduction if deer had not been reaching epidemic proportions. Hypotheses and assumptions can be put to the test. Thus, understandings of long-term ecological change derived from models or chronosequence studies, can be checked against reality. Assumptions of stability can also be tested. Thus, in Lady Park Wood it has been possible to demonstrate the fluctuations in the balance between beech, ash and lime, and thereby to recognise that there is no single 'correct' structure and composition to which this (or any other) woodland reserve should aim.

Scientific standards of long-term studies

Records inherited from the initiators of long-term studies are generally incomplete in some important sense, and they rarely conform to modern design and statistical standards. Analyses are thus often restricted, and editors may not be inclined to make allowances. More fundamentally, long-term studies may be criticised for lacking an hypothesis. This happened to us when we submitted a 50-year study of demographic change in a population of wych elm (Peterken & Mountford 1998). My reply was that I did not know if the initiator had an hypothesis about elm, and that there was no sign that he anticipated the 1970s elm disease outbreak. We could have written the account as if there had been an hypothesis, but that would have misrepresented the study.

An increasing flow of long-term studies indicates that perceptions may be improving. More ecologists seem to appreciate the need to observe what actually happens on the ground, which enables us to test models of development and assumptions about conservation objectives and methods.

Assumptions and experience in the design of long-term studies

Those who today are designing studies that they intend to become long-term will probably assume (i) that clear objectives are required, (ii) that these objectives will be sustained in the long-term to generate predictable research outcomes, (iii) that the design must meet the highest scientific standards, (iv) that detailed records must be made, (v) that regular, frequent recording must be planned and achieved, (vi) therefore, that funding will be regular and predictable, and (vii) that databases must be electronic.

These assumptions can be challenged. Experience from our stand studies shows that objectives will probably change, that outcomes are unpredictable, that simple, practical design layouts may last longer than statistically-correct layouts, that funding is irregular, that regular recording may be less cost-effective than opportunistic recording, and that computer databases may be less durable and accessible than paper records. Clear objectives, correct design and precise plans may be necessary to gain initial approval and funding, but the realistic researcher should see these more as pretexts than predictions

Of course, other kinds of long-term study may be predictable and plannable. Massive undertakings, such as the Hubbard Brook studies (Bormann & Likens 1989), achieve a higher standard of design and record. However, the general experience of other kinds of long-term study is that simplicity is the key, combined with sustained interest by the initiator (Taylor 1989).

A broad view of long-term studies

It is worth recognising that long-term studies may emerge from observations that were not intended to be long-term. Thus, some existing long-term studies started as a single survey, with no intention of repeating the observations. They became long-term because (i) the record survived, (ii) the observations could be relocated on the ground, and (iii) someone was interested in repeating the observations. For example, the plot recorded by Worth in 1921 to describe Wistmans Wood was relocated and re-recorded by Proctor *et al.* (1980) and thus became a long-term study.

Extending this idea, it is worth recognising old photographs as informal long-term studies. Photographs preserve unedited information, which may provide good qualitative information on unforeseen developments. Repeat photographs and sustained fixed-point photography (eg Gruell *et al*. 1982) may afford a greater general appreciation of change than careful, quantitative records.

Organisation of long-term studies

Long-term studies tend to change hands, from initiator to successors, and from one institution to another. For example, the Denny transect has been the responsibility of Southampton University (Botany department), is now largely the responsibility of Ed Mountford and myself, and may soon become the responsibility of Forest Enterprise as part of its monitoring programme for the New Forest. This, however, can lead to problems of "ownership". For example, the plots in the Black Wood of Rannoch, which were started by the Forestry Commission in 1948, were saved by local staff of the Nature Conservancy Council through recording and re-marking in 1984, and extended by new transects established in 1986. In the 1990s, the Forestry Commission resumed interest, but just when Ed Mountford and myself obtained an EU grant under which we would have re-recorded both the plots and the transects. A similar accident has recently happened with a 1961 plot established by the Nature Conservancy in Coed Cymerau NNR.

Clearly, there is a case for some co-ordination. Ideally, perhaps, a single archive should be established as an Ecological Archive Office. Failing that, it would be valuable to establish a clearing-house, which holds information on which projects exist, where the record is held, and who is for the time-being responsible.

The value of an Ecological Archive Office as a simple repository should not be underestimated. Almost every study that has become long-term has survived a period of apathy. This usually happens between the time when the initiator loses interest (say, 10 years), and the time when changes become interesting (say, 25 years). Furthermore, ecologists die, and executors will probably not be ecologists. Even those who appoint specialist executors only half solve the problem. For example, I hold records bequeathed by Eustace Jones (Lady Park Wood and other sites) and Molly Spooner (Wistmans Wood), but the records survive only because I remain interested. I still hold many box files on 1960s records originally made by Richard Steele and others. But for me, they would have been thrown out when he retired as Director-General of NCC in the 1980s. We need somewhere to place material which may be of long-term value.

Conclusions

No doubt this view of long-term studies in woodland is unduly coloured by the particular kind of study with which I have been involved. Perhaps studies of other elements of woodlands, such as invertebrates or physical attributes, can effectively become long-term rather sooner than 25 years. However, even here one must distinguish between short-term fluctuations and long-term trends, and be prepared to observe unexpected changes (eg Likens 1985). If we value long-term observations and wish to increase the chance that new studies will become long-term, we must adopt an approach which is alien to short-term researchers, by (i) designing them around human limitations, and (ii) paying as much attention to archiving and the institutional memory as to ecology and research design.

References

BACKMEROFF, C.E. & PETERKEN, G.F. 1989. Long-term changes in the woodlands of Clairinsh, Loch Lomond. *Transactions of the Botanical Society of Edinburgh*, **45**, 253-297.

BORMANN, F.H. & LIKENS, G.E. 1981. *Patterns and process in a forested ecosystem.* New York, Springer.

CRAMPTON, A.B., STUTTER, A.B., KIRBY, K.J. & WELCH, R.C. 1998. Changes in the composition of Monks Wood National Nature Reserve (Cambridgeshire) 1964-1996. *Arboricultural Journal*, **22**, 229-245.

GRUELL, G.E., SCHMIDT, W.C., ARNO, S.F. & REICH, W.J. 1982. *Seventy years of vegetation change in managed Poderosa pine forest in western Montana.* USDA Forest Service, General Technical Report, INT-130.

LIKENS, G.E. 1985. An experimental approach for the study of ecosystems. *Journal of Ecology*, **73**, 381-396.

MOUNTFORD, E.P., PETERKEN, G.F. & BURTON, D. 1998. Long term monitoring and management of Langley Wood. Peterborough, *English Nature Research Reports*, No. 302.

MOUNTFORD, E.P., PETERKEN, G.F., EDWARDS, P.E. & MANNERS, J.G. 1999. Long-term change in growth, mortality and regeneration of trees in Denny Wood, and old-growth wood-pasture in the New Forest (UK). *Perspectives in Plant Ecology, Evolution and Systematics*, **2**, 223-272.

PETERKEN, G.F. 1981. *Woodland conservation and management.* London, Chapman and Hall. (2nd edition 1993)

PETERKEN, G.F. & BACKMEROFF, C. 1988. *Long-term monitoring in unmanaged woodland nature reserves*. Peterborough, Nature Conservancy Council (Research and Survey in Nature Conservation 9).

PETERKEN, G.F. & JONES, EW. 1987. Forty years of change in Lady Park Wood: the old growth stands. *Journal of Ecology*, **75**, 477-512.

PETERKEN, G.F. & MOUNTFORD, E.P. 1996. Effects of drought on beech in Lady Park Wood, an unmanaged mixed deciduous woodland. *Forestry*, **69**, 117-128.

PETERKEN, G.F. & MOUNTFORD, E.P. 1998. Long-term change in an unmanaged population of wych elm subjected to Dutch elm disease. *Journal of Ecology*, **86**, 205-218.

PROCTOR, M.C.F., SPOONER, G.M & SPOONER, M. 1980. Changes in Wistman's Wood, Devon: photographic and other evidence. *Report and Transactions of the Devonshire Association for the Advancement of Science, Literature and Art*, **112**, 43-79.

STRAYER, D., GLITZENSTEIN, J.S., JONES, C.G., KOLOSA, J., LIKENS, G.E., McDONNELL, M.J., PARKER, G.G. & PICKETT, S.T.A. 1986. *Long-term ecological studies: an illustrated account of their design, operation and importance to ecology*. New York, Institute of Ecosystem Studies. (Occasional Publications 2).

TAYLOR, L.R. 1989. Objective and experiment in long-term research. *In*: G.E. LIKENS, ed., *Long-term studies in ecology*, pp.20-70. New York, Springer.

LONG-TERM STAND CHANGE IN A NEAR-NATURAL OAKWOOD ON CLAIRINSH ISLAND, SCOTLAND

E.P. Mountford

17 Butler Road, Wem, Shropshire, SY4 5YP (e-mail edmountford@edmountford.freeserve.co.uk)

Summary

Changes in unmanaged semi-natural stands at Clairinsh were quantified by means of two permanent transects recorded in 1961, 1986 and 1998. The island was treated as coppice-with-standards up until 1913, when it was last felled to leave a shelterwood of mainly oak. The composition of the wood remains strongly influenced by pre-1913 treatment, and, to a lesser extent, by continued control of non-native invasive tree species. Nevertheless, during the past 85 years the wood has developed structures and processes that are typical of natural temperate woodland.

In gaps left after the 1913 felling, regeneration established in which seedlings of shade-intolerant trees, especially birch (*Betula* spp.) and shrub species were common. After the gaps infilled, a period of stem-exclusion prevailed and only a few dominant trees remained vigorous. Losses were high amongst the post-1913 recruits, which became overtopped by the promoted canopy oaks *Quercus* spp. Only a few young birch, oak, ash *Fraxinus excelsior* and alder *Alnus glutinosa* managed to sustain vigorous growth and recruit into the canopy. An understorey re-initiation phase began a few decades after the stand had closed. Below the canopy, shade-tolerant holly *Ilex aquifolium* and rowan *Sorbus aucuparia* developed strongly. By 1998 the pattern of growth, mortality and gap creation had shifted. Most of the surviving dominant oaks had peaked in diameter growth and mortality had decreased, becoming increasingly associated with storm-damage. The canopy had started to open up, but gaps remained too small, scattered and ephemeral, to provide substantial opportunities for the regeneration of new light-demanding trees or shrubs. The potential long-term composition of woodland is discussed.

Introduction

Increasing interest is being shown in long-term monitoring in minimum intervention reserves (Peterken 1996; Parviainen *et al.* 1999). In Great Britain, a national programme of studies on long-term change in unmanaged semi-natural woodland based on permanent transects/plots was initiated in the mid-1980s under the general supervision of Dr. George Peterken of the Nature Conservancy Council (Peterken & Backmeroff 1988). Other long-term woodland monitoring studies have since been formalised or initiated (Hall, Kirby & Morecroft 1999). This paper reports on changes following a third recording in one of these minimum-intervention reserves. It follows on from a recording made in 1986 (Backmeroff & Peterken 1989), and extends the monitoring to almost four decades.

Site details

Clairinsh (national grid reference NS 413 899) is a 6ha island located in the southern part of Loch Lomond, Central Scotland (Backmeroff & Peterken 1989). It is part of the Loch Lomond National Nature Reserve and is maintained by Scottish Natural Heritage.

W11 *Quercus petraea-Betula pubescens-Oxalis acetosella* woodland (Rodwell 1991) covers most of the island, with oak, *Quercus petraea*, *Q. robur* and hybrids and a few downy birch *Betula pubescens* growing over an understorey of mainly rowan *Sorbus aucuparia*, holly *Ilex aquifolium* and hazel *Corylus avellana*. A marginal low-lying base-rich area to the north-east of the island has much ash *Fraxinus excelsior*, oak and downy birch, growing over a mixed understorey. At the opposite end of the island, another marginal area with base-poor soils has mostly alder *Alnus glutinosa*, birch, and oak, growing over rowan, holly and hazel. These marginal stands correspond to types of W7 *Alnus glutinosa-Fraxinus excelsior-Lysimachia nemorum* woodland, and towards the shoreline they grade into W2 *Salix cinerea-Betula pubescens-Phragmites australis* woodland (Rodwell 1991).

The island lies within a zone naturally dominated by oak woodland (McVean & Ratcliffe 1962; Birks 1988), but has been influenced by human activity for many centuries (Tittensor 1969, 1970; Placido 1986; Backmeroff & Peterken 1989). Parts have been cleared, but there is no evidence of cultivation (Placido 1986). The wood was managed from at least the 17th century, when coppice-with standards treatment prevailed; oak was generally promoted, and several species were planted in. The island was last coppiced in 1913, when the underwood was cut and a shelterwood of oak was left by retaining standard trees and singling the oak stools. The wood has been left unmanaged since, apart from the felling of some non-native trees during 1970-71 (beech *Fagus sylvatica*, larch *Larix* spp. and rhododendron *Rhododendron ponticum*) and in 1998 (western red cedar *Thuja plicata*).

Recording

Ken Wallace working for the Nature Conservancy made the baseline records for Clairinsh. During 1960-2, he made an assessment of the make up and structure of the wood because it exhibited interesting features, such as mixed age composition and continuing natural regeneration, not found in other oakwoods (Wallace 1962). Two transects were established, each 11 yards (10m) wide, and covering 190 yards (174m) and 396 yards (362m) length. Within them all trees and dense holly groves were mapped, and measurements made of stem girths and heights. In addition, three profile diagrams, each covering three sections length, were drawn for three separate parts of the transects, and twenty oaks were aged by taking stem cores.

Christa Backmeroff, working for the Nature Conservancy Council, undertook a second recording of the transects during June 1986 (Backmeroff 1986). To standardise the recording, new 20m-wide transects were recorded following the approximate centre line of the earlier transects. Virtually all the ground covered by the earlier transects was included, and all trees attaining 1.3m height were mapped and girths measured.

All the ground recorded in the 1986 transects was recorded again during April 1998. Trees attaining 1.3m height were mapped. Girths were measured and each stem was allocated to a canopy layer. In addition, canopy gaps, established seedlings, and areas covered by low-growing holly suckers were mapped.

Analysis

Full details of the precise analysis methods and statistical tests applied are available from the author. The data set provided information on 4154 stems. Change was examined in the combined area of the core and marginal stands and within both the Wallace and Backmeroff transects. The core stand dominates the central free-draining higher part the island, whilst the marginal base-rich and base-poor stands respectively occupy adjacent low-lying areas to the north and south. The Wallace transect covered 0.484ha or 9% of the island area, and the Backmeroff transect covered 0.950ha or 18% of the island area. Comparisons were constrained by different thresholds used at each recording. Statistical procedure followed Zar (1984) and tests were carried out using Microsoft Excel Version 7.0a and Statistica Release 4.5 computer packages. Although most individual stems were confidently re-identified, a few errors, omissions, misidentifications, and problems caused by changes in the recording methodology had to be 'adjusted' for.

Results

Stand development before 1961

Backmeroff and Peterken (1989) give details of stand change to 1961. The core and marginal base-poor stands had closed over and remained intact, whereas the marginal base-rich stand had remained somewhat open. The combined stem size-class distribution in 1961 revealed a scatter of larger trees retained in 1913 and abundant regeneration since.

Oaks dominated the canopy in 1961 (Table 1), largely due to the retention of existing oak standards and coppice stools that were either promoted by singling or simply left uncut in 1913. By 1961 most of these were >10m high, >75cm g.b.h., and in the canopy layer.

Most post-1913 recruits were overtopped and shaded by 1961. Oak recruits included both stems from coppice stools and maidens. Virtually all of these were in the understorey, and coppice individuals had thinned to one or a few stems per stool. Oak regeneration was especially poor in the marginal base-rich stand, despite abundant gaps and seed-bearing standards. Recruitment of birch, mostly from seed, was abundant apart from in the marginal base-rich stands. Very few oak and birch recruits had sustained rapid growth and these were associated with larger gaps. Ash recruitment was largely restricted to the marginal base-rich stand. It was most abundant as coppice growth on the open ground immediately above the shoreline, whilst elsewhere only one maiden had developed into a vigorous tree below a gap. Most alder were apparently from coppice stools and most had recruited into the marginal base-poor stand, where a few large individuals had developed into canopy or sub-canopy trees below an original large gap.

By 1961, a shade-tolerant understorey of rowan and holly had established in the core and marginal base-poor stands (Table 1). Most individuals were small and appeared have recently seeded in under the closed canopy, but a few may have been coppiced or retained in 1913. Hazel formed a minor part of the understorey as scattered individuals and aggregated groups. Several large individuals that appeared to be coppice from 1913 remained in reasonable vigour, but others shaded by oak standards were in decline. A few small individuals may have been recent recruits. In contrast the more open-canopied marginal base-rich stand had a diverse shrub layer. These, ranked in abundance, were hawthorn *Crataegus monogyna*, rose *Rosa* spp., broom *Sarathamnus scoparius*, blackthorn *Prunus spinosa*, guelder rose *Viburnum opulus*, hazel, and rowan. These helped infill some of the gaps in the main stand and formed a low-growing scrub with the ash coppice stools on the ground immediately above the shoreline.

21

Table 1: Density and basal area of large and small individuals in the core and marginal stands in 1961, 1986 and 1998. In the Wallace transect large individuals had the largest stem ≥0cm g.b.h., small individuals had the largest stem <10cm g.b.h. in 1961 and 5-9.5cm g.b.h. in 1986 and 1998; and basal area was based on large individuals and only the largest stem on multi-stemmed individuals. In the Backmeroff transect large individuals had the largest stem ≥10cm g.b.h., small individuals had the largest stem <10cm g.b.h., and basal area was based on all stems ≥5cm g.b.h.

| | Large individuals (n ha^{-1}) | | | | | Small individuals (n ha^{-1}) | | | | | Basal area (m² ha^{-1}) | | | | |
| | Wallace transect | | | Backmeroff transect | | Wallace transect | | | Backmeroff transect | | Wallace transect | | | Backmeroff transect | |
	1961	1986	1998	1986	1998	1961	1986	1998	1986	1998	1961	1986	1998	1986	1998
Oak	227	190	178	178	163	10	12	4	11	4	17.9	23.5	25.6	22.5	24.0
Birch	207	122	101	157	131	33	12	6	14	7	2.1	2.0	2.1	2.5	2.6
Rowan	140	211	227	194	194	110	81	50	72	59	0.6	1.0	1.4	1.5	1.8
Hazel	76	33	39	31	39	31	41	48	93	96	0.2	0.1	0.1	0.3	0.4
Holly	72	322	364	310	350	233	244	186	256	233	0.5	1.4	2.1	2.0	3.0
Ash	37	31	31	31	29	2	4	-	4	-	0.7	0.4	0.5	0.6	0.7
Alder	33	21	12	16	12	-	2	-	1	-	0.5	0.5	0.3	0.6	0.5
Hawthorn	14	17	10	13	12	4	2	2	11	6	0.2	0.1	<0.1	0.2	0.1
Beech	2	2	2	2	2	-	-	2	-	3	<0.1	<0.1	<0.1	<0.1	<0.1
Blackthorn	2	12	19	21	25	8	10	19	75	83	<0.1	<0.1	<0.1	0.1	0.1
Larch	2	-	-	-	-	-	-	-	-	-	0.3	-	-	-	-
Pine	2	2	2	1	1	-	-	-	-	-	0.1	0.3	0.5	0.2	0.2
Yew	-	6	6	8	9	6	-	4	2	4	-	<0.1	0.1	<0.1	0.1
Cedar	-	2	2	1	1	-	-	-	-	-	-	<0.1	<0.1	<0.1	<0.1
Guelder rose	-	2	-	1	-	10	-	4	9	22	-	<0.1	-	<0.1	<0.1
Rose	-	-	2	-	1	14	10	19	25	20	-	<0.1	<0.1	<0.1	<0.1
Crab apple	-	-	-	2	2	-	-	-	-	-	-	-	-	<0.1	<0.1
Willow	-	-	-	1	-	-	-	-	-	-	-	-	-	<0.1	-
Bird cherry	-	-	-	-	-	-	2	-	2	2	-	-	-	<0.1	<0.1
Broom	-	-	-	-	-	25	-	-	2	9	-	-	-	<0.1	-
Total	769	930	950	928	936	473	409	320	490	444	23.1	29.3	32.7	30.4	33.5

22

Table 2: Mortality, reduction and recruitment of large individuals in the core and marginal stands between 1961-86 and 1986-98. Large individuals had the largest stem ≥10cm g.b.h.

	Wallace transect										Backmeroff transect			
	1961		1961-86		1986	1986-98					1986	1986-98		
	No. alive	No. died	No. reduced to <10cm g.b.h.	No. recruited & <10cm g.b.h. or not recorded in 1961	No. alive	No. died	No. reduced to <10cm g.b.h.	No. recruited & ≥10cm g.b.h. in 1961 & <10cm g.b.h. in 1986	No. recruited & <10cm g.b.h. in 1961 & 1986	No. recruited & <10cm g.b.h. or not recorded in 1986	No. alive	No. died	No. reduced to <10cm g.b.h.	No. recruited & <10cm g.b.h. or not recorded in 1986
Oak	110	21	-	3	92	6	-	-	-	-	169	13	-	-
Birch	100	44	2	5	59	10	-	-	-	-	149	25	-	-
Rowan	68	10	2	46	102	1	3	-	8	4	184	11	4	15
Hazel	37	14	10	3	16	1	1	3	2	-	29	1	2	11
Holly	35	1	1	123	156	3	6	-	3	26	294	5	6	49
Ash	18	3	-	-	15	1	-	-	-	1	29	3	-	2
Alder	16	5	1	-	10	4	-	-	-	-	15	4	-	-
Hawthorn	7	2	-	2	7	3	-	-	-	-	1	4	-	4
Beech	1	1	-	1	1	-	-	-	-	-	2	-	-	-
Larch	1	1	-	-	-	-	-	-	-	-	-	-	-	-
Pine	1	-	-	-	1	-	-	-	-	-	1	-	-	-
Blackthorn	1	-	1	6	6	-	1	-	-	4	20	3	1	8
Yew	-	-	-	3	3	-	-	-	-	-	8	-	-	1
Crab apple	-	-	-	1	1	1	-	-	-	-	3	1	-	-
Cedar	-	-	-	1	1	-	-	-	-	-	1	-	-	-
Rose	-	-	-	-	-	-	-	-	1	-	-	-	-	1
Guelder rose	-	-	-	1	1	-	1	-	-	-	1	-	1	-
Willow	-	-	-	-	-	-	-	-	-	-	1	1	-	-
Total	395	102	17	195	471	30	12	3	14	36	917	72	14	91

23

Table 3: Mortality, growth and recruitment of small individuals in the core and marginal stands between 1961-86 and 1986-98. Small individuals in the Wallace transect had the largest stem <10cm g.b.h. in 1961 and 5-9.5cm g.b.h. in 1986 and 1998. Small individuals in the Backmeroff transect had the largest stem <10cm g.b.h.

| | Wallace transect | | | | | | | | | | | | Backmeroff transect | | | | |
| | 1961 | 1961-86 | | | | | 1986 | 1986-98 | | | | | 1986 | 1986-98 | | | |
	No. alive	No. died	No. grown to ≥10cm g.b.h.	No. reduced to <5cm g.b.h.	No recruited & ≤10cm g.b.h. in 1961	No. recruited & <5cm g.b.h. or not recorded in 1961	No. alive	No. died	No. grown to ≥10cm g.b.h.	No. reduced to <5cm g.b.h.	No. recruited & ≥10cm g.b.h. in 1986	No. recruited & <5cm g.b.h. or not recorded in 1986	No. alive	No. died	No. grown to ≥10cm g.b.h.	No. recruited & ≥10cm g.b.h. in 1986	No. recruited & not recorded in 1986
Holly	113	1	102	1	1	108	118	19	29	-	5	2	243	32	49	6	54
Rowan	53	4	30	0	1	19	49	4	12	1	2	-	68	15	15	4	14
Hazel	18	2	3	5	8	6	22	1	5	2	1	9	88	15	11	2	27
Birch	16	9	5	-	2	3	7	4	-	-	-	1	13	7	-	-	1
Broom	12	10	-	1	-	-	1	1	-	-	-	4	2	2	-	-	0
Rose	7	3	-	-	-	1	5	-	1	-	-	5	24	12	1	-	8
Oak	5	3	2	-	-	6	6	4	-	-	-	-	1-	6	-	-	-
Guelder rose	5	5	-	-	-	-	-	-	-	-	1	1	9	4	-	1	15
Blackthorn	4	1	2	-	1	3	5	-	2	-	1	5	71	18	7	1	32
Yew	3	-	3	-	-	-	-	-	-	-	-	2	2	-	1	-	2
Ash	2	2	-	-	-	2	2	1	1	-	-	-	4	2	2	-	-
Hawthorn	2	-	2	-	-	1	1	-	1	-	-	1	10	2	4	-	2
Alder	-	-	-	-	-	-	1	1	-	-	-	-	1	1	-	-	-
Beech	-	-	-	-	-	-	-	-	-	-	-	1	-	-	-	1	2
Bird cherry	-	-	-	-	-	-	-	-	-	-	-	-	2	-	-	-	-
Total	240	40	149	7	14	149	207	35	51	3	10	51	547	116	90	15	166

Stand development from 1961 to 1998

Tables 2 and 3 show the number and basal area of large and small individuals in the core and marginal stands in 1961, 1986 and 1998. Stand development generally matched that reconstructed for the previous half century, and mostly reflected continued undisturbed growth. There was a substantial increase in basal area to 33.5m² ha⁻¹, an increase in large individuals, but a decline in small individuals. Some canopy disturbance did occur and canopy gaps covered 19% of transect area in 1998. Gaps were mostly limited in extent and had been caused by windthrow (from a hurricane in 1968), branch snap, and crown deterioration of canopy oaks. The marginal base-rich stand had most canopy gaps at 32% of the transect area because, in addition to windthrown and windsnapped canopy trees, some areas had failed to close over since the 1913 fellings.

Oak remained dominant throughout as its basal area steadily increased. Recruitment was very limited. Large oak individuals suffered low mortality ($m = 0.6$-0.8% yr⁻¹ over the two periods in the two transects), but small post-1913 recruits suffered higher loss. Most mortality was linked to the exclusion of smaller-girth trees, while storms killed and damaged several of the other larger oaks. Three died after been windthrown in a hurricane in 1968, and another was windsnapped during the 1986-98 period. Records of other trees windblown in 1968 showed that the western side of the island suffered most (Figure 2 in Backmeroff and Peterken 1989), and that only a very few trees survived the fall and remained alive in 1998. Otherwise, a few dominant trees in the core woodland appeared to have deteriorated for no clear reason.

Based on observations of surviving trees made in 1998, many oaks suffered crown damage during the hurricane and in later storms. This ranged from major crown snapping to minor branch loss, and the consequent damage to understorey stems hit by falling trees and debris. In total 32 out of 126 canopy trees had some damage recorded. A few trees had or were replacing lost crown through sprout growth, the most vigorous of which came from directly below break points.

Surviving oaks grew at very different rates. Trees that ended up overtopped in the understorey and sub-canopy layers grew significantly less (average = 0.190 cm g.b.h. yr⁻¹) than canopy trees (average = 0.775 cm g.b.h. yr⁻¹) (ANOVA $F = 31.4$, $p<0.001$). The overtopped trees included most of the small-girth post-1913 recruits, but a few of these maintained above average growth and grew into the canopy layer; one, which was located below a large gap in the marginal base-rich stand in 1961, was the fastest-growing individual throughout.

Birches declined overall, despite a slight increase in basal area. Recruitment was minimal, although a few saplings developed in large gaps created by the 1968 hurricane. The density of both large and small individuals fell, as the post-1913 recruits thinned heavily. Mortality of large individuals was relatively high ($m = 1.6$-2.3% yr⁻¹ over the two periods in the two transects), but the population of small suppressed individuals suffered even more ($m = 3.3$-7.1% yr⁻¹). Most losses were linked to exclusion by dominant trees, but storm-damage killed two out of three larger-girth trees, with the 1968 hurricane throwing a 132cm g.b.h. tree over, and a 64 cm g.b.h. tree recorded as dead in 1986 had had its top snapped off. Only a few surviving trees had suffered storm-related damage.

Growth of surviving birch was very variable, but most surviving canopy trees grew at least moderately. Most of the small-girth post-1913 recruits, which had become overtopped in the understorey and sub-canopy by 1998, grew slowly. Nevertheless, a few of these maintained above average growth and grew into the canopy. Several overtopped trees showed signs of growth release following the creation of gaps in the overstorey, although girth growth in these remained far below that of most canopy trees

Ash survived only in the marginal base-rich stand. The surviving post-1913 ash recruits remained moderately abundant and, despite the loss of the large standard that died standing and fell with its roots rotted by 1987, basal area increased. Alder was scarce in the core and marginal base-rich stands and declined to very low levels, and even in the marginal base-poor stand, where it had been most abundant, it declined greatly. Exclusion was the main cause of loss.

During 1961-98, rowan continued to develop strongly in the core stand, moderately in the marginal base-poor stands, but remained sparse in the marginal base-rich stand. Recruitment of slow-growing small seedlings far exceeded mortality which was low, especially for large individuals (m<0.6% yr^{-1} over the two periods in the two transects). Many of the 40 individuals that died were apparently excluded, but a substantial number were hit by canopy debris or falling trees, and the largest tree to die was windthrown in the 1968 hurricane.

Holly continued to develop strongly in the understorey in the core and marginal base-poor stands, but by 1998 only a single small individual had recruited into the marginal base-rich stand. The sudden increase in numbers between 1961-86 was probably partly because the 1961 population was underestimated. In addition there was low mortality of large individuals (m<0.2% yr^{-1} over the two periods in the two transects), slow growth of small individuals, and steady recruitment of new individuals.

Hazel remained a minor understorey species in the core and marginal base-poor stands. Large individuals declined greatly before 1986 mainly due to shading: mortality was high (m = 1.9% yr^{-1}) and on survivors, especially those with large girth stems, there was a reduction in the total number of live stems. After this the population stabilised as mortality and reduction dropped, and several small individuals grew on. The population remained dominated by small-sized individuals with a high turnover, but several stools appeared to be growing reasonably well in 1998 below the broken canopy and almost 50 new subsidiary stems had grown to ≥5cm g.b.h. during 1986-98. In contrast, hazels in the marginal base-rich stand developed and remained abundant. No individuals died, a few new seedlings recruited, and some exceptionally large stools developed.

In the core stand nine other minor species were recorded. Larch and beech were felled as exotic trees, but the single Scots pine *Pinus sylvestris* was retained. By 1986 two nearby beech had grown vigorously to replace a felled beech, and by 1996 two beech saplings had established elsewhere. Yews *Taxus baccata* formed a minor but persistent part of the understorey and some recruitment after 1986 increased them. A few hawthorn, crab apple *Malus sylvestris* and new brooms were present in 1998, but rose had died out. In the marginal base-poor stand, hawthorn and guelder rose died out, but a single vigorous beech sapling recruited.

In the marginal base-rich stand shrub species remained prominent. Blackthorn increased to become the most numerous species, forming a weak understorey below gaps in the main stand. Rose remained abundant as a climber or lateral sprawler growing in the stands immediately above the shoreline. Guelder rose became quite abundant, but hawthorn declined slightly. Otherwise, a few crab apple, bird cherry *Prunus padus* and new broom were present in 1998. A western red cedar that recruited and grew vigorously was felled just after the 1998 recording.

Discussion

The three detailed recordings at Clairinsh provide a basis to reconstruct past-stand change, and to extrapolate to longer-term changes in such mesotrophic oak woodland.

In general the patterns of natural stand development parallel those described by Oliver (1981) and Oliver and Larson (1996) for North American temperate forests. In the gaps left after 1913, a wave of regeneration established with shade-intolerant trees, especially birch and shrub species, common. Once the gaps filled, a period of stem-exclusion prevailed. The most vigorous trees survived and grew on to increase the stand basal area, whilst the weakest were thinned out. Losses were highest amongst the post-1913 recruits that soon became overtopped by the promoted canopy oaks. Nevertheless, a few birch, oak, ash and alder recruits managed to sustain vigorous growth and grew into the canopy. An understorey re-initiation phase began a few decades after the stand had closed: shade-tolerant holly and rowan, and a few yew developed strongly below the 10-20m high canopy layer of semi-mature oaks, although the overall distribution was distinctly patchy and most strongly developed in the core woodland. These episodes of gap-phase regeneration and understorey re-initiation increased the age-structure of the wood and produced a quasi-negative exponential size-distribution, typical of continuously regenerating woodland (Peterken 1996).

By 1998 the stands had entered the old-growth phase of forest development. Basal area levels had reached levels comparable to natural old-growth ranges of 25-35m² ha⁻¹ (Parker 1989) and 21-42m² ha⁻¹ (Martin 1992) for North American and 37m² ha⁻¹ in stands in central Europe (Falinski 1986). Large trees had become more frequent (7% of oaks attained 200cm g.b.h.), and, moreover, the pattern of growth, mortality and gap creation had shifted. The surviving dominant old oaks appeared to have peaked in diameter growth at about 100cm g.b.h., and thereafter had grown at a much reduced rate in old-age. Instead of high mortality rates caused primarily by competition between individual trees, mortality had dropped and canopy trees had become increasingly affected by storm-damage and, to a lesser extent, unidentified physiological disorders, reflecting; (i) reduced competition between the crowns of dominant trees that become less vigorous in old-age (Oliver & Larson 1996); and (ii) increased susceptibility of large-sized, ageing trees to physical damage and physiological disorders (Mueller-Dombois 1986, 1987; Franklin *et al.* 1987).

Gap creation has remained relatively small-scale and moderately episodic. The loss of a few dominant canopy trees has created openings up to 0.02ha, mostly following an intense gap creation episode when a hurricane struck in 1968, but this has been followed by gradual, more-or-less annual, small-scale crown loss. Although marginal oak crowns have been generally slow to expand into gaps, several damaged trees have reformed crowns by producing vigorous epicormic shoots from below the breakage points. Light conditions below gaps have been further moderated by the presence of sub-gap growth in the form of suppressed trees, shade-tolerant understorey shrubs, and, rarely, toppled trees that have survived and developed new low-growing crowns. Although the patchy break-up of the canopy has increased the structural diversity of the main stands, it has yet to provide substantial opportunities for the regeneration of new shade-intolerant trees or shrubs. So far just one recorded group of birch and rowan saplings has grown well on the root mound of an oak toppled in the 1968 hurricane. Indeed, despite uprooting several large canopy trees, this last event has surprisingly had only a limited effect on the composition of the wood.

Despite the elimination of light-demanding shrubs from most of core and marginal base-poor stands, this layer has been conserved at the northern-end of the island in the marginal base-rich and associated shoreline shingle stands. Here, the stand has remained relatively open and well-lit. Large oaks are less abundant and the stands close to the shoreline are low-growing and composed mainly of light-crowned ash. Rather than advancing towards old-growth, these stands have tended to perpetuate themselves as young-growth and as such have remained the focus of tree and shrub diversity on the island.

The future structure and composition of Clairinsh

Predicting long-term stand change in natural woodland is problematic, not least at Clairinsh because the present stands are not even one generation removed from past-management. In addition natural woodland is subject to unpredictable disturbance events that can profoundly alter the course of stand development (e.g. Peterken & Mountford 1996).

At Clairinsh, the main form of natural disturbance has been storms. Damaging storms are a regular periodic occurrence across Scotland (Andersen 1953; Allen 1992), but they are rarely widespread catastrophic events and result mostly in localised damage (Andersen 1953; Steven & Carlisle 1959; Peterken & Stace 1989; Peterken 1996). The January 1968 storm caused far more damage to the neighbouring island of Inchcailloch (unpublished data), whereas Clairinsh appears to have escaped with little damage both from this and from severe storms in the early 1950s (Andersen 1953). Other disturbances, caused by boat visitors and reserve management, have had minor effects on the stand and promoted no unexpected regeneration.

Assuming this type of disturbance regime continues, and that the island remains free of grazing (it is naturally isolated against livestock and deer grazing, with only a few red deer seen swimming out in temporary stopovers), it is possible to broadly outline the likely future composition.

In the short-term oak looks set to remain dominant for many more decades: the canopy trees have the potential to live for several centuries (Jones 1959) and some have been able to redevelop lost crown from epicormic growth. Until the next catastrophic storm occurs, further gradual break-up of the oak canopy is likely, as the ageing oaks continue to suffer storm-damage and become less able to refill the gaps. Opportunities for regeneration appear to depend on the scale of the next catastrophic storm: oak is most likely to require large-scale openings for regeneration, with vegetative stump and trunk resprouting likely to form an important part of this (e.g. Thomas et al. 1994; unpublished data from Toy's Hill, Kent).

The shade-tolerant holly, rowan and yew understorey should continue to develop below the oak canopy. Further gradual canopy break-up should further release rowan and holly and encourage the growth of hazel in the understorey. Such development will reduce opportunities for gap-phase regeneration of birch, oak, and other shade-intolerant species.

Birch looks set to continue to decline as suppressed trees are excluded and larger canopy trees reach natural maturity (Harding 1981). Nevertheless, a few younger trees appear able to sustain themselves in the old-growth stands, being released into growth below small canopy openings, an attribute more often associated with shade-tolerant trees (Emborg 1998). Despite birch having regenerated best at Clairinsh in large-scale openings, a little regeneration did occur in windthrow gaps created in 1968. Birch therefore looks able to survive at low levels for many decades, and when the next catastrophic storm occurs it could form a major part of the seedling regeneration.

Long-term survival of more light-demanding trees and shrubs is most likely in the northern parts of the marginal base-rich and shoreline stands where the structure is more open and the soils are more suitable for their growth.

It appears that, without direct control, beech would infiltrate the woodland: it has, albeit in low numbers, seeded and grown into canopy trees in the core woodland and, beyond the transects, established a population of pole-sized trees at the southern-most end of the island. Similarly, sycamore appears capable of such infiltration: it presently occurs as scattered trees and suppressed saplings around the shoreline, but in the northern-most part of the marginal base-rich stand there are several vigorous saplings in open areas to the east of the permanent transect.

These findings concur with the summary given by Backmeroff and Peterken (1989) who considered that; (i) birch and oak would eventually decline to small populations; (ii) holly, rowan and yew and perhaps hazel would become more abundant in shade-tolerant understorey layer; (iii) alder would become a minor component in the marginal woodland; and (iv) beech and sycamore would spread and might become dominant unless controlled by felling.

Acknowledgements

This paper and the field recording in 1998 are part of the RENFORS project - Regeneration of Native Forest Stands for timber production and environmental value - funded by the European Communities Directorate-General for Agriculture (Contract FAIR PL95-0420). The study at Clairinsh has relied on the input of Ken Wallace (Nature Conservancy), George Peterken and Christa Backmeroff (Nature Conservancy Council), and several staff from Scottish Natural Heritage.

References

ALLEN, J.R.L. 1992. Trees and their response to wind: mid Flandrian strong winds, Severn Estuary and inner Bristol Channel, southwest Britain. *Philosophical Transactions of the Royal Society of London B*, **338**, 335-364.

ANDERSEN, K.F. 1953. Gales and gale damage to forests, with special reference to the effects of the storm of 31st January 1953, in the north-east of Scotland. *Forestry*, **27**, 97-121.

BACKMEROFF, C.E. 1986. *Loch Lomond NNR: long-term changes in the woodland of Clairinsh.* Unpublished Report, Peterborough, Nature Conservancy Council.

BACKMEROFF, C.E. & PETERKEN, G.F. 1989. Long-term changes in the woodland of Clairinsh. *Transactions of the Botanical Society of Edinburgh*, **45**, 253-297.

BIRKS, H.J.B. 1988. Long-term ecological change in the British uplands. *In*: Usher, M.B and Thompson, D.B.A., eds. *Ecological change in the uplands*. Oxford, Blackwell Scientific, 37-56.

EMBORG, J. 1998. *Suppression and release during canopy recruitment of* Fagus sylvatica *and* Fraxinus excelsior. Unpublished manuscript.

FALINSKI, J.B. 1986. *Vegetation dynamics in temperate lowland primeval forests*. Dordrecht, Geobotany 8, Junk.

FRANKLIN, J.F., SHUGART, H.H. & HARMON, M.E. 1987. Tree death as an ecological process. *BioScience, 37*, 550-556.

HALL, J.E., KIRBY, K.J. & MORECROFT, M.D. 1999. *Minimum intervention woodlands and their use for ecological research in Great Britain*. JNCC Report No. 295. Peterborough, Joint Nature Conservation Committee.

HARDING, J.S. 1981. Regeneration of birch (*Betula pendula* Enrh. and *Betula pubescens* Roth.). Addendum to: Newbold, A.J. and Goldsmith, F.B. *Discussion Papers in Conservation No. 33*, University College London, pp 83-112.

JONES, E.W. 1959. Biological Flora of the British Isles: *Quercus* L. *Journal of Ecology*, **47**, 169-222.

MARTIN, W.H. 1992. Characteristics of old-growth mixed mesophytic forests. *Natural Areas Journal,* **12**, 127-135.

McVEAN, D.N. & RATCLIFFE, D.A. 1962. *Plant communities of the Scottish Highlands*. London, HMSO.

MUELLER-DOMBOIS, D. 1986. Perspectives for an etiology of stand-level dieback. *Annual Review of Ecological Systematics*, **17**, 221-243.

MUELLER-DOMBOIS, D. 1987. Natural dieback in forests. *BioScience, 37*, 569-583.

OLIVER, C.D. 1981. Forest development in North America following major disturbances. *Forest Ecology and Management*, **3**, 153-168.

OLIVER, C.D. & LARSEN, B.C. 1996. *Forest stand dynamics*. Cambridge, Cambridge University Press.

PARKER, G.R. 1989. Old-growth forests of the central hardwood region. *Natural Areas Journal,* **9**, 5-11.

PARVIAINEN, J., LITTLE, D., DOYLE, M., O'SULLIVAN, A., KETTUNEN, M. & KORHONEN, M. (eds.) 1999. *Research in forest reserves and natural forests in European countries - Country reports for the COST Action E4: forest reserves research network*. Joensuu, European Forest Institute Proceedings No.16. pp 267-294.

PETERKEN, G.F. 1996. *Natural woodland*. Cambridge, Cambridge University Press.

PETERKEN, G.F. AND BACKMEROFF, C.E. 1988. *Long-term monitoring in unmanaged woodland nature reserves*. Research & Survey in Nature Conservation No.9. Peterborough, Nature Conservancy Council.

PETERKEN, G.F. & MOUNTFORD, E.P. 1996. Effect of drought on beech in Lady Park Wood, an unmanaged mixed deciduous woodland. *Forestry,* **69**, 117-128.

PETERKEN, G.F. & STACE, C.E. 1987. Stand development in the Black Wood of Rannoch. *Scottish Forestry,* **41**, 29-44.

PLACIDO, C. 1986. *Loch Lomond NNR reserve record; part 1 (preliminary).* Unpublished report, Nature Conservancy Council South West (Scotland) Region.

RODWELL, J.S. 1991. *British plant communities. Volume 1: woodlands and scrub.* Cambridge, Cambridge University Press.

STEVEN, H.M. & CARLISLE, A. 1959. *The native pinewoods of Scotland.* Edinburgh and London, Oliver and Boyd.

THOMAS, R.C., KIRBY, K.J. & COOKE, R.J. 1994. The fate of storm-damaged trees in Ham Street National Nature Reserve, Kent. *In*: KIRBY, K.J. & BUCKLEY, G.P., eds., *Ecological responses to the 1987 Great Storm in the woods of south-east England.* (English Nature Science No 23.) Peterborough, English Nature. pp 65-80.

TITTENSOR, R.M. 1969. *The role of man and other ecological factors in determining the variation within oakwoods east of Loch Lomond.* Unpublished M.Sc. Thesis. Edinburgh, University of Edinburgh.

TITTENSOR, R.M. 1970. History of the Loch Lomond oakwoods. *Scottish Forestry,* **24**, 100-118.

WALLACE, K. 1962. *Clairinsh Nature Reserve: observations on the woodland contents.* Unpublished report. Edinburgh, Nature Conservancy.

ZAR, J.H. 1984. *Biostatistical Analysis* (2nd Edition). London, Prentice-Hall International.

LONG TERM STRUCTURE AND VEGETATION CHANGES IN A NATIVE PINEWOOD RESERVE IN NORTHERN SCOTLAND

C. Edwards and **W.L. Mason**
Silviculture North Branch, Northern Research Station, Roslin, Midlothian, EH25 9SY

Summary

Long term changes in stand structure and vegetation composition were studied in four large permanent plots in the Black Wood of Rannoch, a native pinewood in Northern Scotland. *Pteridium aquilinum* and *Juncus* spp. were recorded in fewer quadrats in 1993 than in 1948. *Vaccinium* spp increased in abundance in one plot, but decreased significantly in all other plots. No major changes were observed in the abundance of *Calluna vulgaris*. Tree recruitment was greater than mortality when averaged over all plots for the 45 year study period. Light levels between plots suggest some have more, or larger, canopy gaps than others. The changes in vegetation abundance and increased tree recruitment may reflect recovery of the pinewood after wartime fellings.

Introduction

The Black Wood of Rannoch is a native Scots pine reserve of 800 ha situated on the southern shores of Loch Rannoch in the Tummel Valley, Perthshire (OS Grid ref: NN 570560). It has a northerly aspect and mean elevation of 250m ASL. Mean annual precipitation is 1300mm yr^{-1} with an accumulated temperature of 1110-875 day-degrees above 5°C. Basic geology is undifferentiated gneisses and schists of Dalradian series, with deep layers of glacial drift material and granite erratics. Soils are predominately podsols on morainic knolls with poorly flushed peaty gleys between the knolls.

Plot History

Ownership of the Black Wood has changed periodically down through the centuries with fellings and disturbance occurring regularly from the seventeenth century. Whayman (1953), Steven and Carlisle (1959) and Arkle and Nixon (1996) have described how former owners systematically felled and removed many trees and allowed overgrazing by domestic and feral stock (principally sheep and deer), preventing natural regeneration. The Black Wood of Rannoch has predominately been a deer forest, although it has been fenced on occasion to keep deer out as well as in. In 1940/41 the Canadian Forestry Corps carried out an extensive felling programme, cutting over much of the wood in the area, removing only the best trees and leaving very small or rough trees (Wonders 1996).

Acquisition of the Black Wood of Rannoch in 1946 (Forest Enterprise 1995) presented the Forestry Commission with a unique opportunity to monitor long term changes in a native pinewood, under conditions of 'non-intervention' management after a period of disturbance and disruption. It was proposed to study and describe the regeneration process in seven permanent plots, located in areas of the wood where the vegetation community was typified by different vegetation species or species mix. Each plot was approximately 0.81 ha in area. Centuries of mismanagement and exploitation however, had left the Black Wood of Rannoch in a moribund and degenerate state. Some authors argued there was inadequate regeneration (Peterken & Stace

1987) and some foresters thought the wood was not capable of regenerating or returning to a natural state again (Brown 1956).

This paper presents results from monitoring changes in vegetation and stand structure over a 45-year period. Interpretation of the changes over the monitoring period is also discussed, and some of the problems encountered with long term monitoring schemes identified.

Methods

Of the seven original unfenced plots established in 1947 only four remain today (Plots 4, 5, 6 and 7). Plots 1-3 were located outwith the native reserve and were converted to plantation in 1956 (Arkle & Nixon 1996). The remaining four plots were assessed in 1948, 1956, 1983/4 and 1993/94 when broad vegetational changes and tree and seedling establishment or mortality were recorded. Several events have disrupted three of the four plots since establishment. An experiment was established over part of plot 5 in 1949, and an area of 0.25 ha fenced. A fence was also erected in part of plot 6 in 1972 as part of an ITE experiment. An electricity power line was located in plot 7 in the same year resulting in all trees under the line and 30 m either side of it being felled. Both events precluded further data recording in 0.17 ha and 0.25 ha respectively from plots 6 and 7. In all assessments and analyses, these areas have been omitted, unless stated otherwise.

Vegetation mapping

Vegetation was mapped on four occasions in each plot; in 1948, 1956 1984 and 1993. The initial mapping was completed by T. C. Mitchell in late 1948 using a Forestry Commission system that combined five main species in thirty different ways according to their dominance or frequency. The main species identified were *Calluna vulgaris* (L.) Hull., *Pteridium aquilinum* (L.) Kuhn., *Molinia caerulea* (L.) Moench., *Vaccinium vitis-idaea* L. and *V myrtillus* L., *Juncus acutiflorus* Ehrh. Ex Hoffm. and *J. conglomeratus* L. In August 1956, D.W. Henman used three levels of association (dominant, sub-dominant, subsidiary) to combine the dominant species representative of each of the five vegetation types identified in 1948. Although these three levels of association did not correspond to any accepted scale of ecological vegetation description at the time, they did allow associations of up to three species to be distinguished and recorded in map form. The 1948 maps were retraced and converted to the new scheme for continuity in 1960.

A period of 27 years passed before the plots were re-mapped in 1984 by students supervised by G.F. Peterken and H. Stace, using the same system of species dominants devised by Henman. Accurate re-location of outer plot boundaries proved difficult where marker posts had rotted and decomposed. Staff and students from Forest Research completed the most recent mapping in 1993. For the purposes of this paper, only the change between 1948 and 1993 was analysed. It has been assumed any change in vegetation frequency between these dates would reflect plot differences due to structural changes over the same period.

For each plot in 1948 and 1993 a grid system was superimposed on to each map, using tracing paper marked out in 180 quadrats. Each vegetation species was recorded as present or absent. This gave a frequency abundance value for all species. These data were then analysed using a chi-squared distribution to compare predicted change with actual change recorded.

Structural changes

The location and species of trees within each plot were marked during 1948 and 1956 using an internal grid system to estimate where the position of each tree was in relation to its neighbour. Peterken and Stace (1987) collected tree position and seedling and sapling location during a repeat exercise in 1984. In 1994, Forest Research re-surveyed all four plots. Tree height, diameter at breast height (dbh), crown depth and crown width in two directions were recorded. Trees >10 cm dbh were aged by taking increment cores. All tree positions were accurately recorded using a theodolite and standard survey techniques. Seedling (<10 cm dbh, <1.3 m in height) positions and species were also recorded.

Each plot map was scrutinised by grid square and trees identified as present, dead or missing, or as new recruits since the previous mapping exercise. The accurate re-location of every individual tree proved difficult. Many discrepancies arose, and mistakes between mapping exercises became clear when all four sets of data were analysed. Mortality and recruitment rates were calculated for each plot for both pine and birch. In cases where it was not possible to rectify discrepancies using available data, the simplest explanation for change using the least number of arguments or assumptions was used.

Light transmittance

A hand held one-metre long ceptometer, the SunScan from Delta T, was used to measure Photosynthetically-active Photon Flux Density (PPFD) (shortwave radiation in 400 -700nm range, expressed in μmol m^{-2} s^{-1}), at sample points located systematically on a 10m x 10m grid system under each permanent plot. At each sample point a mean of 64 light measurements along the one-metre length were recorded. Each value was compared with an above canopy reference value and expressed as the percentage of available above canopy light transmitted through the canopy.

Results

Vegetation

In all four plots *Juncus* spp. and *P. aquilinum* in plots 5, 6 and 7 were recorded in fewer quadrats in 1993 than in 1948. This change was significant for *Juncus* spp in all plots, and for plots 5 and 6 for *P. aquilinum* (p<0.001) (Table 1).

Table 1. Percentage abundance of plant species mapped in 1948 and 1993 in the Black Wood of Rannoch. Values indicate the percentage of quadrats in which each species was recorded. Changes in abundance are not significant unless indicated by * = P<0.05, **= P<0.01, ***= P<0.001

Species	Plot 4		Plot 5		Plot 6		Plot 7	
	1948	1993	1948	1993	1948	1993	1948	1993
P. aquilinum	10.6	11.7	97.1	74.6***	99.3	33.8***	20	17.6
C. vulgaris	100	100	96.4	93.5	99.3	98	86.4	99.2***
M. caerulea	28.3	21.7	4.3	2.2	10.6	1.3***	-	27.2***
V. myrtillus & *V. vitis-idaea*	78.9	88.9**	99.3	81.2***	99.3	78.8***	67.2	48.8**
J. acutiflorus & *J. conglomeratus*	28.9	3.3***	4.3	0.0*	9.9	0.0***	56	26.4***
Grasses	0.0	1.1	-	-	-	-	-	-

In both plots 5 and 6, all species were recorded with lower abundance in 1993 than 1948. This was significant in plot 6 for *P. aquilinum, M. caerulea, Vaccinium,* spp. and *Juncus* spp. (all P<0.001) but not for *C. vulgaris*. The abundance of *C. vulgaris* did not change significantly between recording intervals in plots 4,5, & 6 but increased significantly in plot 7 (P<0.001). *Vaccinium* spp increased in abundance in plot 4 (P<0.001) between recording intervals, but decreased significantly in all other plots.

Structure

Plot 4 had the highest basal area and mean tree age of all plots (Table 2), and was the only plot where mortality rates were higher than recruitment (Table 4). In general, recruitment of both Scots pine and birch exceeded mortality between successive assessments (Figure 1), resulting in substantial increases in tree population size since 1948. There are substantial quantities of new seedlings recorded in all plots (Table 3) which represent a large potential population of new cohorts.

Table 2. Tree species data from the four permanent assessment plots in the Black Wood of Rannoch (excluding areas disturbed by fences and power lines).

	N° Scots pine trees in 1994 (>1.3 m ht; >0.10 cm dbh)	Basal Area (m² ha⁻¹) all spp	Mean tree		Age range
			Ht (m)	Age	
Plot 4	130	27.3	15.7	119	23-263
Plot 5	58	21.2	14.4	80	13-204
Plot 6	81	14.9	11.0	37	9-140
Plot 7	114	25.7	14.3	57	14-153

Note: 1. Basal area calculation includes birch, rowan and pine trees.
        2. Age is at 1.0 m ht above ground level.

Table 3. Total number of tree seedlings (<1.3 m height) recorded in all plots, 1994. (Plot areas adjusted to exclude fenced enclosures but not disturbed area in plot 7).

Species	Number	Density (ha⁻¹)
Pine	427	150
Birch	50	17.6
Rowan	11	3.9
Willow	3	1

Table 4. Recruitment and mortality rates expressed as numbers per hectare per year for the 45-year period between 1948 and 1994. A positive difference indicates that recruitment is exceeding mortality.

	Plot 4	Plot 5	Plot 6	Plot 7
Recruitment (SP and Birch)	0.87	2.00	3.52	3.28
Mortality (SP and Birch)	1.13	1.21	1.29	0.67
Difference	-0.26	0.79	2.23	2.61

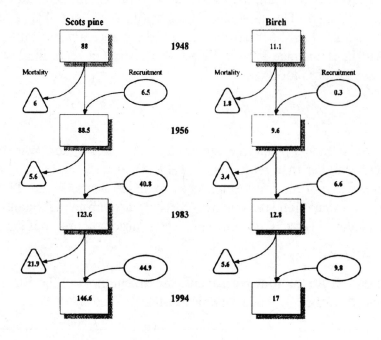

Figure 1. Mortality and recruitment rates for Scots pine and birch in all plots (1948-1994). Values indicate mean number of trees ha^{-1} present on each assessment date, and the mortality and recruitment rates between each successive assessment date.

Light transmittance

Mean values of percentage transmitted light under canopy relative to above canopy light levels, are lower in plot 4 (28.8%) than plot 5 (37.6%) or plot 6 (40.8%), although these differences are not statistically significant (Figure 2).

Discussion

Few native pinewoods in northern Scotland have the classic J-curve distribution of tree ages expected of forests with constant recruitment and mortality (Oliver & Larson 1996). Many show group recruitment after disturbance events (Goucher & Nixon 1996; Parker & Peet 1984). The four plots in this study are no exception (Figure 3); they indicate punctuated small-scale gap recruitment probably as a result of past management interventions. Typically this is identified in the age distribution by high frequency of trees in some classes and very low frequency in others.

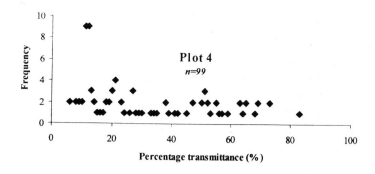

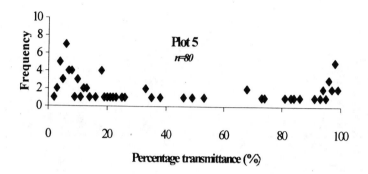

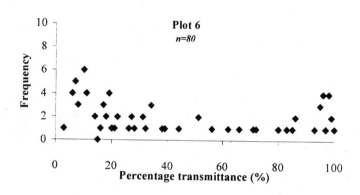

Figure 2. Light transmittance frequency scatter plot for three permanent plots in the Black Wood of Rannoch. Values indicate the percentage of light above the canopy that is transmitted to beneath the canopy. Low transmittance values are associated with increasing canopy cover.

In plots 5, 6 and 7 tree recruitment is exceeding mortality. As a consequence there is a higher frequency of individuals in the youngest age classes. Recruitment is probably occurring because stands are recovering from disturbance during felling operations in the period 1940-1942. Either advance regeneration was released by the felling operation, or new seedling germination occurred, or both. Many individuals assessed in 1994 were <20 years old and substantial numbers of seedlings, too small to accurately age, were also present, although the disturbance event occurred 50 years earlier. Either seedlings require this length of time to grow to 10 cm dbh, or there was a delay following felling before conditions under the stand became suitable for recruitment. In parts of the Black Wood reserve, seedlings have taken an average 16 years to reach 1 m height in good growing conditions (Edwards unpublished) it is conceivable they would take considerably longer under less favourable conditions.

After the 1956 assessment, Brown (1956) gave a gloomy prediction for the pine wood, which was not producing many new seedlings and still looked moribund and degenerate. Recruitment and mortality data for each interval between assessments suggests that not until after the 1956 assessment is there a substantial increase in recruitment. Seedfall patterns in the late 50s to early 60s indicate substantial seed production was occurring in the plots (McIntosh & Henman 1981). Abundant seed (i.e. 3 million seeds ha⁻¹) was recorded twice in a 5 year period, with germination percentages of 37%-100% of full seed.

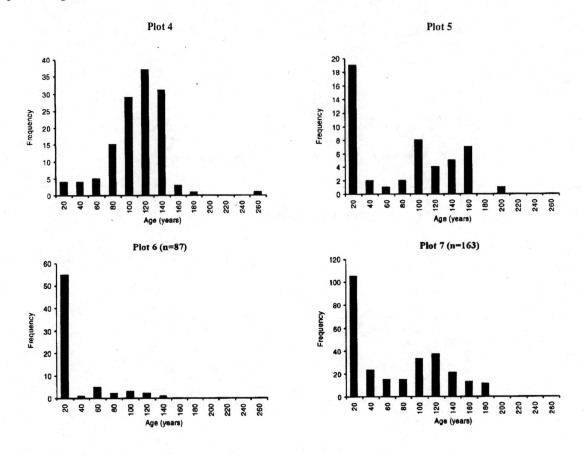

Figure 3. Age class distribution for pine and birch >10cm dbh in the four permanent assessment plots, the Black Wood of Rannoch.

The age distribution pattern for plot 4 differs significantly from the other plots in the study (Figure 3). There is a peak of individuals in the 120 years age class with lower frequency in all younger age-classes. In addition, plot 4 has a greater basal area than any of the other plots (Table 2) and is the only plot where mortality exceeds recruitment (Table 4). This suggests that the plot has passed a phase of recruitment and may now be passing into a new phase of stem exclusion.

In all plots there is a large pool of seedlings too small to be counted in the recruitment/mortality assessments. These seedlings constitute a reserve of potential cohorts ready to take advantage of any new canopy disturbance. This can be seen clearest in plot 7, under the disturbed electrical line, where conditions are now most favourable for seedling germination and establishment.

Økland and Eilertsen (1996) have shown that understorey pine vegetation abundance may vary from year-to- year, making single year studies separated by long intervals difficult to interpret. Changes may be due to long-term trends or yearly variation due to short term factors. When the permanent plots were established at the Black Wood of Rannoch, vegetation mapping proceeded without thought to the statistics to be used in analysis of results. The data collected has thus proved difficult to analysise because recording vegetation presence by mapping is subjective and has not generated interval data for analysis. Interpreting our results, based on two assessments, therefore requires caution. However, it is likely that broad overall changes and the direction in which the vegetation composition is changing are indicated, although detailed specific information about each species is limited.

Changes in stand structure may also explain long term vegetation changes. The increase in V*accinium* spp frequency is recorded only in plot 4, where the stand appears to be approaching maximum canopy cover. This is indicated by no transmittance values >80%. Transmittance values >80% are only possible if the canopy has large gaps allowing light to penetrate without attenuation and such gaps are by contrast apparent in plots 5 & 6. Reduced frequency of high transmittance values in plot 4 suggests a more homogeneous canopy cover, which may increase shading thus favouring the shade-tolerant *Vaccinium* spp over other species (Figure 2). In all other plots, disturbance events of the 1940s would have opened the canopy to favour light demanders, especially C*alluna*. In plots 5 and 6 the large gaps in the canopy continue to allow relatively high light levels to be maintained. Concurrent with increasing tree recruitment there will be a corresponding decrease in available soil moisture, through increased evapotranspiration and interception by the canopy (Pyatt & Craven 1979). This will act against the moisture requiring species such as *Juncus* spp. and *Molinia*.

Bracken may well have benefited initially from disturbance, but has reduced in frequency in all plots with increasing recovery of the stand. The dense bracken stand in the first few years following felling may also have acted as a barrier to seedling recruitment and may partially explain the 20 year delay in seedling recruitment. It is likely *Calluna* would have also increased following disturbance, and has remained abundant while conditions have been favourable.

Conclusions

The Black Wood of Rannoch is recovering from the most recent intervention event with population and stand structural development following an apparent 30 year time lag. Information on structural changes during this period has been collected at appropriate intervals. Vegetational changes were more difficult to interpret due to the long time period between assessments. However, we need to make assessments at intervals at an appropriate time-scale to pick-up changes in structural development and vegetational fluctuations. Long term monitoring need not necessarily mean long time intervals between measurements, but more the sustained commitment to recording at intervals suited to answer the basic questions underlining the monitoring objectives.

Acknowledgements

The authors thank students Paul Arkle, David Brown and G Neal for collection of field data, and David Anderson who helped supervise Earthwatch volunteers. The co-operation and support of staff from Forest Enterprise is also appreciated. Jonathan Humphrey commented on an earlier version of this manuscript.

References

ARKLE, P.J. & NIXON, C.J. 1996. Structure and growth characteristics of Scots pine *Pinus sylvestris* L. in long-term monitoring plots within the Black Wood of Rannoch native pinewood. *Scottish Forestry,* **50**, 145-150.

BROWN, J.M.B. 1956. *A visit to the Black Wood of Rannoch.* Forestry Commission Research Internal report (Unpublished).

FOREST ENTERPRISE. 1995. *Tay Forest District: Black Wood of Rannoch Management Plan, 1 April 1995 - 31 March 2005.* Unpublished Draft.

GOUCHER, T. & NIXON, C.J. 1996. A study of age structure in three native pinewoods in Lochaber. *Scottish Forestry*, **50,** 17-21.

McINTOSH, R. & HENMAN, D.W. 1981. Seedfall in the Black Wood of Rannoch. *Scottish Forestry*, **35**, 249-255.

ØKLAND, R.H. & EILERTSEN, O. 1996. Dynamics of understorey vegetation in an old-growth boreal coniferous forest, 1988-1993. *Journal of Vegetation Science*, **7**, 747-762.

OLIVER, C.D. & LARSON, B.C. 1996. *Forest stand dynamics update edition.* USA, John Wiley & Sons, Inc.

PARKER, A.J. & PEET, R.K. 1984. Size and age structure of conifer forests. *Ecology*, **65**, 685-689.

PETERKEN, G.F. & STACE, H. 1987. Stand development in the Black Wood of Rannoch. *Scottish Forestry,* **41**, 29-44.

PYATT, D.G. & CRAVEN, M.W. 1979. Soil changes under even-aged plantations. *In:* FORD, E.D., MALCOLM, D.C. & ATTERSON, J., eds. *The Ecology of even-aged forest plantations.* pp 369-386. Edinburgh, Institute of Terrestrial Ecology.

SUNSCAN MANUAL. 1997. *Canopy Analysis System Technical Manual.* Cambridge, Delta-T Devices Ltd.

STEVEN, H.M. & CARLISLE, A. 1959. *The native pinewoods of Scotland.* Edinburgh and London, Oliver and Boyd.

WHAYMAN, A. 1953. The Black Wood of Rannoch. *Scottish Forestry*, **7**, 112-117.

WONDERS, W.C. 1996. The Canadian Forestry Corps in Scotland during WWII. *Scottish Forestry*, **50**, 85-92.

A CENTURY OF VEGETATION CHANGE AT BROADBALK WILDERNESS

Gary Kerr[1], Ralph Harmer[1] and Stephen R. Moss[2]

[1] Forestry Commission Research Agency, Alice Holt Lodge, Wrecclesham, Farnham, Surrey, GU10 4LH, UK.

[2] IACR - Rothamsted, Harpenden, Hertfordshire, AL5 2JQ, UK.

Summary

Broadbalk Wilderness contains a small area of secondary woodland which developed following the abandonment of a wheat field in 1882. Archived and published data from the area has provided the best available information on the long-term dynamics of natural colonisation at a single site in Britain. However, despite continuity of ownership and a good system of data recording and retrieval, the area has lacked a clear objective and a common assessment protocol. Hence the area has fragmented and it has been difficult to piece together exactly what has happened. Future projects which involve long-term monitoring of woodlands must ensure they meet basic criteria, otherwise resources could be wasted.

Introduction

The Continuous Wheat Experiment (CWE) is one of the classic long-term studies at IACR Rothamsted (Johnston 1994) and was established by Sir John Lawes in 1843. Broadbalk Wilderness is an area which was formed following a decision by Lawes to make the area of each of the plots in the CWE equal by cutting off a portion of the experimental field. This presented Lawes with an opportunity to study how the wheat crop would perform if it was not harvested and meticulously hand weeded as the rest of the CWE. In 1882 he told the 39th wheat crop just before the harvest of 1882 "I am going to withdraw all protection from you, and you must for the future make your own seedbed and defend yourself in the best way you can against the natives, who will do everything in their power to exterminate you" (Lawes 1884 quoted in Garner 1965). As predicted the area quickly became dominated by couch grass *Agropyron repens* and smaller and smaller amounts of self-sown wheat were noted in 1883, 1884, 1885 until there was none in 1886. The area has survived but is now divided into three areas: (a) an undisturbed area which has colonised to mixed woodland, (b) an area where invading woody plants have been removed ('grubbed') for over 90 years, and (c) an area created in 1957, when half the grubbed area was changed to a regime of regular grazing by sheep (Figure 1). The objective of this paper is to summarise the methods used to monitor the area during the past 117 years and to learn lessons which may be useful in designing future long-term monitoring studies in woodlands.

Description of study site

The Wilderness (Figure 1) is a small area (0.26 ha) situated 700 m northwest of the main building at IACR - Rothamsted near Harpenden, Hertfordshire (National grid reference TL121137). It is on a gently undulating plateau at an elevation of about 130 m. The underlying geology is chalk. The soil is classified as a leached brown soil with a loamy surface layer overlying clay-with-flints; it is moderately well drained with a pH of 7.8 (Jenkinson 1971). The soil contains between 3 and 5% free $CaCO_3$ which is a residue of heavy liming of the site before 1843; this has probably buffered any significant change in pH over time (Blake 1994).

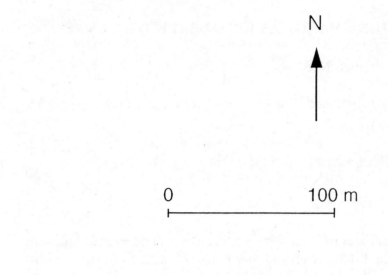

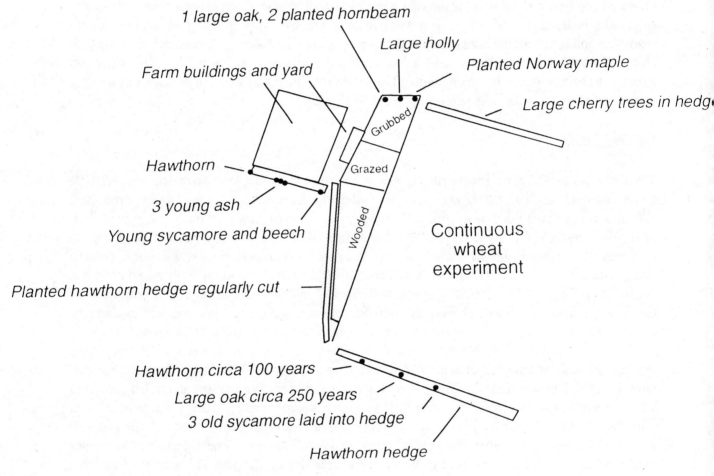

N

0   100 m

1 large oak, 2 planted hornbeam

Large holly

Planted Norway maple

Farm buildings and yard

Large cherry trees in hedge

Grubbed

Hawthorn

Grazed

3 young ash

Wooded

Young sycamore and beech

Continuous
wheat
experiment

Planted hawthorn hedge regularly cut

Hawthorn circa 100 years

Large oak circa 250 years

3 old sycamore laid into hedge

Hawthorn hedge

Figure 1. Layout of the Broadbalk Wilderness

Monitoring of Broadbalk Wilderness

Table 1 describes how Broadbalk Wilderness has been monitored in the period 1886 to 1999.

Table 1. Chronology of monitoring at Broadbalk Wilderness

Date and surveyor	Method	Results and comments	Woody species recorded in wooded area
1886 27 July J.J. Willis	List of species on a 9 point DAFORS scale (which spanned exceedingly abundant to rare).	40 herbaceous species were recorded. No formal experiment plan and neither methods nor surrounding vegetation were described. Published in Lawes (1895)	none
1894 Autumn J.J. Willis	List of species on a 9 point DAFORS scale (exceedingly abundant to rare).	51 herbaceous and 5 woody species recorded. Methods were not described and the presence of the woody species was only noted in the text. Published in Lawes (1895)	*Corylus avellana, Crataegus monogyna, Fraxinus excelsior, Quercus robur, Rosa sp.*
1903 3 April and 3 July J.J. Willis	List of species on a 9 point DAFORS scale (exceedingly abundant to rare).	56 herbaceous and 10 woody species recorded. Methods again not described. Published in Brenchley and Adam (1915) who omitted records of woody species.	*Corylus avellana, Crataegus monogyna, F. excelsior, Hedera helix, Ligustrum vulgare, Lonicera periclymenum, Prunus avium, Q. robur, Rosa sp, Rubus fruticosus.*
190?	Half of the area was grubbed.	No record of the date the work was carried out. Reasons for the change were not documented.	n/a
1913 Several dates in summer Brenchley and Adam	**Wooded area**: species list **Grubbed area**: list of species on 9 point DAFORS scale.	10 herbaceous and 5 woody species recorded. No estimate of cover except recording the area as an oak-hazel wood in Brenchley and Adam (1915). 75 herbaceous species Published by Brenchley and Adam (1915) which compared results with earlier information.	*Corylus avellana, H. helix, Q. robur, Rosa* spp., *R. fruticosus.* n/a
1938 May Clapham and Baker	**Wooded area**: list of species with a DAFORS scale for woody plants only.	13 herbaceous and 15 woody species recorded. Published in Tansley (1939)	*Acer campestre, A. pseudoplatanus, Cornus sanguinea, Corylus avellana, Crataegus monogyna, F. excelsior, H. helix, Ilex aquifolium, Ligustrum vulgare, P. spinosa, Q. robur, Rosa* spp., *Rubus fruticosus, Salix caprea, Sambucus nigra.*

Date and surveyor	Method	Results and comments	Woody species recorded in wooded area
1944 November Thurston	**Grubbed area**: listed woody species to be grubbed.	Written description with some indication of abundance.	n/a
1945 March Thurston	**Wooded area**: list of species and diameter of some trees and shrubs at 0.91 m above ground level.	5 herbaceous and 16 woody species. Records well documented. Written description of vegetation with indication of distribution and abundance but not all trees and shrubs were measured. Briefly mentioned in Witts (1965) and data referred to in Kerr, Harmer and Moss (1996).	*A. campestre, A. pseudoplatanus, Cornus sanguinea, Corylus avellana, Crataegus monogyna, F. excelsior, H. helix, I. aquifolium, L. vulgare, Lonicera periclymenum, P. avium, P. spinosa, Q. robur, Rosa spp., Rubus fruticosus, S. nigra.*
1951 29 August Curtis	**Grubbed area**: list of species	35 herbaceous and 11 woody species. Unpublished but briefly described by Witts (1965).	n/a
1957	The Field Plots Committee decided to convert half of the grubbed area to a grazing regime.	This was to investigate the changes in herbage induced by stocking. Initially mown but grazing began in March 1960 (Garner 1965).	n/a
1957-65 10 dates in 1956/57 and in subsequent years 1 day in March/April. Thurston	**Wooded area** Not surveyed	In December 1959 six trees were felled as they were shading the continuous wheat experiment (Witts 1965).	n/a
	Grubbed/grazed area 17 permanent quadrats, species lists and % cover.	49 herbaceous and 10 woody species initially present. Five quadrats were abandoned as being atypical. Vegetation of ungrazed part remained unchanged whereas grazed area showed changes in species composition. Briefly described in Witts (1965).	n/a
1960 4 April Witts	**Wooded area**: girth of all 'large' trees measured at 0.91 m above ground level.	10 woody species noted, no definition of 'large'. Some description of vegetation on site. Referred to briefly by Witts (1965) and in Kerr, Harmer and Moss (1996).	*A. campestre, A. pseudoplatanus, Corylus avellana, Crataegus monogyna, F. excelsior, H. helix, I. aquifolium, P. spinosa, Q. robur, S. nigra.*

Date and surveyor	Method	Results and comments	Woody species recorded in wooded area
1969 Jenkinson	**Wooded area**: girth of trees and shrubs >2.5 cm diameter at 1.3 m above ground level.	Only seven woody species measured. Only trees ≥5 m from woodland edge measured. Carried out as part of detailed study of soils, results published in Jenkinson (1971).	*A. campestre, A. pseudoplatanus, Corylus avellana, Crataegus monogyna, F. excelsior, Q. robur, S. nigra.*
1995-9 Harmer, Kerr and Peterken 16/17 August 1995 6+18 May 1998	**Wooded area** Species list, cover estimated by Braun-Blanquet scale. Height, dbh and position of all over- and understorey trees > 2.5 cm at 1.3 m. Species list by extensive walking survey.	20 herbaceous and 17 woody species recorded. Flora of nearby Knott Wood and local hedgerow recorded (the hedgerow was laid in winter 1998/9). Some tree information published in Kerr, Harmer and Moss (1996).	*A. campestre, A. pseudoplatanus, Cornus sanguinea, Corylus avellana, Crataegus monogyna, F. excelsior, H. helix, I. aquifolium, Ligustrum vulgare, P. spinosa, P. avium, Q. robur, Ribes uva-crispa, Rosa sp., Rubus fruticosus, S. nigra, Taxus baccata.*
6 May 1998 and 7 April 1999	**Grubbed/grazed area** Casual observations of species present	16 dicotyledons observed, many were woodland species or weeds that can tolerate shade.	

45

Conclusions

The information collected and archived about Broadbalk Wilderness is the best available data for studying natural colonisation over a long period of time at a single site. In terms of contemporary interest in natural colonisation as a method of woodland creation, the site is a useful demonstration of its advantages and disadvantages (Kerr, Harmer & Moss 1996).

The survival of the site has been assured by being under a single ownership and there have not been any development pressures which have led to loss of area. However, the erection of the farm buildings in 1913 and its subsequent development into a major operational focus may have disturbed the hydrology and/or the nutrient balance of the site.

The site owner has been primarily focused on agricultural research and has a culture of record keeping, information storage and, importantly, the ability to retrieve it.

The motivation for the establishment of the area was the desire of Lawes to rationalise the CWE and observe the performance of untended wheat against invading weeds. As stated objectives for the area they were quickly achieved. However, we have been unable to find any amended statement of objectives for the area since and the lack of an objective has had the following effects:

i.   Fragmentation of the area: the grubbed and grazed areas were formed for reasons of contemporary research interest at the time the changes were made. The reasons for these changes, and others, have not been well documented.

ii.   Lack of a long-term systematic plan for assessing the area: periodic assessments have used different methods and reflect the interests of the surveyors. Subsequently it has been difficult to piece together exactly what has happened and when it occurred.

Some difficulties have arisen, which are not unique to Broadbalk, due to the use of common names for plants and changes in botanical nomenclature.

Using the experience of Broadbalk Wilderness it is possible to identify important factors for successful long-term monitoring of woodlands.

i.   The existence of the woodland must be assured and there must be an ability to control changes to the surrounding area.

ii.   There must be a clearly defined objective for the study.

iii.   The assessment protocol must be clearly thought out in terms of the objective and the practicalities of carrying it out over a long period of time. Subsequent assessments must use the same protocol, or if change is deemed necessary, in a way that does not prevent direct comparison with previous assessments.

iv.   Information must be clearly recorded and safely stored so that it can be easily retrieved.

References

BLAKE, L. 1994. *Changes with time in the chemistry of soil acidifying under woodland and grassland.* Reading, University of Reading, PhD Thesis.

BRENCHLEY, W.E. & ADAM, H. 1915. Recolonisation of cultivated land allowed to revert to natural conditions. *Journal of Ecology*, **3**, 193-210.

GARNER, H.V. 1965. Broadbalk Wilderness History. Harpenden, *Annual Report for the Rothamsted Research Station (1964)*, pp. 218-219.

JENKINSON, D.S. 1971. The accumulation of organic matter in soil left uncultivated. Harpenden, *Annual Report for the Rothamsted Research Station (1970)*, pp. 113-137.

JOHNSTON, A.E. 1994. The Rothamsted Classical Experiments. *In*: R.A. LEIGH and A.E. JOHNSTON, eds. *Long-term experiments in Agricultural and Ecological Sciences*, pp 9-37. Wallingford, CAB International.

KERR, G. , HARMER, R. & MOSS, S. 1996. Natural colonisation: a study of Broadbalk Wilderness. *Aspects of Applied Biology* 44, *Vegetation Management in Forestry, Amenity and Conservation Areas: Managing for Multiple Objectives,* pp 25-32.

LAWES, J.B. 1884. In the sweat of thy face shalt thou eat bread. *Country Gentleman*, 18 September.

LAWES, J.B. 1895. Upon some properties of soils. *Agricultural Students Gazette*, **7**, 64-72.

TANSLEY, A.G. 1939. *The British Islands and their Vegetation.* Cambridge, Cambridge University Press.

WITTS, K.J. 1965. Broadbalk Wilderness Flora. Harpenden, *Annual Report for the Rothamsted Research Station* (1964), pp. 219-222.

THE COLONISATION OF GROUND FLORA SPECIES WITHIN A 38 YEAR OLD SELF-SOWN WOODLAND

Kevin Walker & Tim Sparks
NERC Institute of Terrestrial Ecology, Monks Wood, Abbots Ripton, Huntingdon, Cambridgeshire PE17 2LS

Introduction

The Monks Wood Wilderness (Figure 1) is a self-sown woodland which developed on a former barley field adjacent to Monks Wood National Nature Reserve in Cambridgeshire. This 4 hectare field, which was probably cleared in Roman times, is known to have been cultivated continuously from at least 1820 to 1960 when it was abandoned and allowed to revert naturally to woodland (Mellanby 1967). Since then a self-sown woodland of mainly *Quercus robur* and *Fraxinus excelsior* has developed with an understorey of *Prunus spinosa*, *Cornus sanguinea* and *Crataegus monogyna/laevigata*. Given the proximity of ancient woodland, which surrounds the wood on three sides, the Wilderness offers an ideal opportunity to examine the colonisation and establishment of woodland species, under near optimal conditions. This short paper summarises the results of a survey of the ground flora 38 years after reversion.

Method

In 1998 the Wilderness was divided into 20 by 20 metre permanently marked plots. These were then further divided into 10 by 10 m grid cells (n = 408), within which the position of tree and shrub species were mapped and ground flora species recorded.

Results

89 ground flora species were recorded during the survey, as well as saplings of 10 species of tree and shrub. 16 typical woodland herbs were recorded of which *Arum maculatum* (Figure 2), *Carex sylvatica* and *Rumex sanguineus* were the most abundant. *Ajuga reptans*, *Geum urbanum*, *Mercurialis perennis* and *Moehringia trinerva*, all of which are uncommon in the surrounding wood, occurred in small numbers and were largely confined to the edges of the Wilderness (Figure 2). The majority of the remaining species are more typical of woodland glade/edge and hedgerows in the vicinity. These include *Glechoma hederacea* and a number of coarse grasses, most notably *Brachypodium sylvaticum*, *Calamagrostis epigejos* and *Carex pendula* (Figure 2) which have increased markedly within the surrounding woodland in recent years (Pollard *et al.* 1998) apparently due to selective grazing by Muntjac deer (*Muntiacus reevesi*) (Cooke 1994).

Discussion

The results of this study show that the establishment of self-sown woodland on ex-arable land can be effective where local seed sources are plentiful (i.e adjacent to ancient woodland). Not surprisingly, colonisation was most rapid for species with large animal or wind dispersed seeds, particularly shrubs and trees with large fruits such as *Quercus robur, Fraxinus excelsior, Cornus sanguinea, Crataegus* spp., *Rosa* spp. and coarse grasses and sedges. In contrast, a number of herbs present in the surrounding woodland are only now beginning to colonise, 38 years after

1947

1971

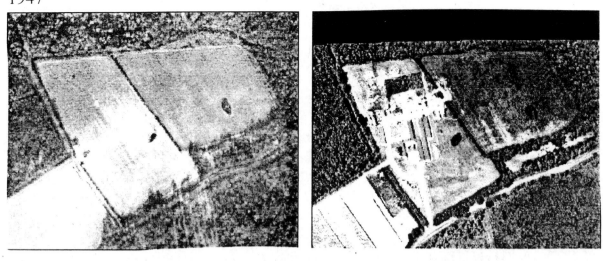

1991

Figure 1. The development of woodland cover in the Wilderness prior to and after the construction of Monks Wood Experimental Station in the early 1960s.

abandonment, and are confined to the edges of the Wilderness. This may be due in part to poor dispersal ability (e.g. *Lamiastrum galeobdonum, Hyacinthoides non-scripta*) and/or declining abundance within the surrounding woodland due to grazing by Muntjac (e.g. *Mercurialis perennis*) (Wells 1994). This deer species was first recorded in the wood in 1972 and has presumably affected the fate of the ground flora and regeneration of tree species (Cooke *et al.* 1995). As a result, less palatable species which are common in the surrounding wood and hedges, such as *Arum maculatum, Calamagrostis epigejos* and *Glechoma hederacea*, are now particularly abundant. These results illustrate the efficacy of self-sowing (and non-intervention management) as a means of restoring woodland where seed-sources are plentiful.

References

COOKE, A.S. (1994). Colonisation by muntjac deer *Muntiaca reevesi* and their impact on vegetation. *In*: D. MASSEY and R.C. WELCH, eds. *Monks Wood National Nature Reserve: the Experience of 40 Years 1953-93*. pp 45-61. Peterborough, English Nature.

COOKE, A.S., FARRELL, L., KIRBY, K. & THOMAS, R. 1995. Change in abundance and size of dog's mercury apparently associated with grazing by muntjac. *Deer*, **9**, 429-433.

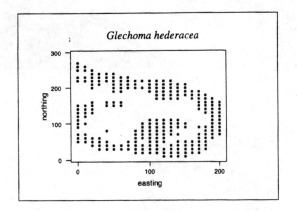

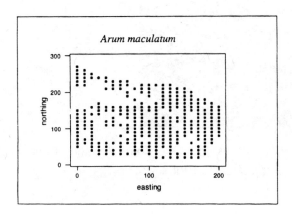

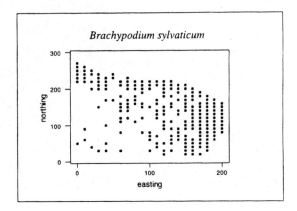

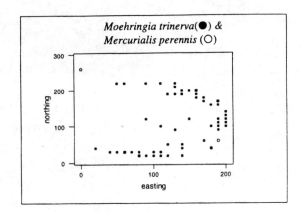

Figure 2. Distribution of typical woodland species within the Wilderness (each circle represents one 10 × 10 m plot).

MELLANBY, K. 1967. The effects of some mammals and birds on the regeneration of oak. *Journal of Ecology,* **5,** 359-366.

POLLARD, E., WOIWOD, I.P., GREATOREX-DAVIES, J.N., YAYES, T.J. & WELCH, R.C. 1998. The spread of coarse grasses and changes in numbers of Lepidoptera in a woodland nature reserve. *Biological Conservation,* **84,** 17-24.

WELLS, T.C.E. 1994. Changes in vegetation and flora. *In*: D. MASSEY and R.C. WELCH, eds. *Monks Wood National Nature Reserve: the Experience of 40 Years 1953-93.* pp 19-25. Peterborough, English Nature.

THE EFFECT OF SHADE AND WEATHER ON WILD DAFFODILS IN WEST DEAN WOODS, WEST SUSSEX

Brian Hopkins
7 Highland Road, Chichester, West Sussex PO19 4QX

Summary

This paper describes the results of monitoring of a population of wild daffodils *Narcissus pseudonarcissus* in a woodland nature reserve from 1979. Shade was the major factor controlling daffodil density. After 12-years, the optimum shade treatment for their conservation had been determined and was put into practice. Shade, unless it was very dense, had little effect on the fraction of bulbs in flower. Flowering was largely controlled by the weather of the previous spring, when warm dry conditions were most favourable.

Introduction

West Dean Woods was created a nature reserve of what is now the Sussex Wildlife Trust in 1975 and was notified as a Site of Special Scientific Interest in 1980. Since its reservation, research and monitoring have been encouraged and there is ongoing, long-term, annual monitoring of several groups: the BTO woodland common bird census since 1975; the ITE butterfly recording survey from 1979; and vascular plants, bryophytes and lichens since 1985. This paper describes the results of monitoring the daffodils *Narcissus pseudonarcissus* since 1979. Nomenclature follows Stace (1991).

One of the four main reasons for making the site a reserve was a colony of wild daffodils, but what is the optimum shade treatment for the conservation of the daffodil colony?

This paper describes an experiment to determine the effects of shade - of both woody plants and of undergrowth - on daffodil density. The effects of weather were also investigated. After twelve years the optimum treatment was clear and was put into practice. The experimental treatments were discontinued, but monitoring continued. The flowering of the daffodils was studied in the same experiment. This is relevant to their conservation both biologically and because their flowering attracts many human visitors.

A fuller account of this work is given in Hopkins (1999).

The site

West Dean Woods Nature Reserve is on the South Downs in West Sussex, where the Upper Chalk rock is covered by up to 1 m of non-calcareous deposits. Woodland, coppice or shrub cover have been continuous since at least 1604 AD (Stewart 1976) and the reserve is species-rich. Much of the area had been managed as coppice-with- standards, although this had been neglected. On reservation, both coppicing and the felling and replanting of oak standards were re-established over most of the reserve. Most standards were oak *Quercus robur*, planted c1870, and most coppice was hazel *Corylus avellana* with some sweet chestnut *Castanea sativa*.

The wild daffodil colony covers 1.65 ha at an altitude of 140-160 m above sea level on an ESE aspect in the north-western corner of the reserve in Compartment 123 (National Grid reference

SU 847159). Coppicing was not re-established here, but the dead fronds of bracken *Pteridium aquilinum* and the tangles of bramble *Rubus fruticosus* agg. have been swiped using a tractor and a 'Bushhog' swipe (an approximately 1.4 m diameter rotary blade at a height of 40-50 cm) during a dry period (when there was little danger of compressing the soil or the bulbs) each winter to allow more light to reach the daffodils.

The mean annual rainfall in West Dean village (4 km SSE and 110 m lower) is 971 mm; in the reserve it is about 1000 mm (Potts & Browne 1983). The nearest, comparable climatic data are for Rustington (24 km SE, 3 m altitude). Allowing for lapse-rate cooling, these give a mean maximum temperature for the hottest month (normally July or August) of about 19 °C and a mean minimum temperature for the coldest month (normally January or February) of about 0 °C.

Methods

Plots and treatments

Five 30 x 20 m plots were selected (Table 1) to give three degrees of woody-plant shade: dense, light and open (virtually none). Two plots under both light shade and open conditions were chosen. One of each was swiped (as already described); the other was left unswiped. Swiping was not possible under dense shade because the closeness of the trees and shrubs prevented the use of a tractor. Care was taken to ensure sufficient differences in cover between the three shade categories and to make the paired plots as similar as possible. Unfortunately the daffodil colony occupied too small an area for the treatments to be replicated.

Table 1: Initial tree and shrub densities and covers of bramble and bracken; densities of mature and flowering daffodil bulbs and increases in these at West Dean Woods

Plot	1	2	3	4	5
Shade treatment	dense	light	light	open	open
Swiped	no	yes	no	yes	no
Woody plants >2cm dbh (1977) (density per sq m)	0.49	0.10	0.10	0.02	0.04
Bramble cover (1979)	light	very sparse	sparse	sparse	light
Bracken cover (1979)	sparse	very sparse	sparse	light	dense
Daffodil density (per sq m) (1979)	98.9	77.2	76.5	10.4	6.1
Density mean increase * (1977-91)	1.5	6.2	2.9	1.9	0.7
Proportional increase‡ (1977-91)	1.5	5.9	3.0	8.1	6.3
Flower density (1979-91) (mean per sq m)	4.7	9.5	8.4	2.5	0.7
Flowers (1979-91 (mean % of mature bulbs)	4.8	9.0	8.8	9.1	7.9

* regression slope; increase per sq m per year
‡ annual increase as a percentage of density

The area covering the experimental plots was about 0.5 ha and contained 18 standards. All were oak and all were large. The dense-shade plot had five times the density of woody plants (>2 cm dbh) on each of the light-shade plots and seventeen times that on the open plots.

The density of undergrowth of both bramble and bracken varied between the plots and bramble appears to have declined during the experiment. The great storm of October 1987 (Kirby & Buckley 1994) blew over some trees on or near the plots and caused an increase in illumination to the edges of Plots 1-3 due to the devastation of a plantation immediately north of the reserve.

Numbers of daffodils

The density of sterile-mature and of flowering daffodil bulbs was determined between late March and late April each year from 1979 to 1987. (The numbers of seedlings and young plants were also recorded, but are not considered in this paper.) Most of the recording was done by trained amateur volunteers under supervision. To confirm the trends, the experiment was continued until 1991 but with the density determinations in alternate years. Following the completion of the experiment, the management was changed but the density of daffodils in the plots continued to be monitored but, from 1995, at five-year intervals. For each assessment, quadrats (25 x 25 cm) were placed at 200 randomly-selected points on each plot. This process was repeated on each plot and in each year with a fresh randomization each time.

Weather

The annual changes in density were compared with weather data for each of the fourteen calendar months from March of the year before recording to April when the records were made. Seven variables were examined for each calendar month: mean maximum temperature; mean minimum temperature; mean temperature (average of the two previous variables); mean soil temperature at 10 cm depth (most bulbs were near the surface); number of days (nights) of ground frost (recorded by a grass-minimum thermometer); total rainfall; and total hours of sunshine. Correlation coefficients were calculated for the mean results on all five plots. When a factor showed significant correlations for two or more consecutive months, additional correlation coefficients were determined for the combined periods.

Density

The initial densities of mature bulbs were very similar on the three shaded plots (77-99 m^{-2}); they were much less on the two open plots (6-10 m^{-2}). Table 1 and Figure 1 show the changes during the experiment and subsequently. (Note the absence of a mean line after yearly counts ceased and the interrupted line for the extrapolation of the regression). During the 12-years, statistically-significant increases in density occurred on all plots ($P < 0.05$ to < 0.001). The only significant differences between the rates on these plots were between Plot 2 (light shade, swiped), which had the highest rate, and Plots 1, 4 and 5 ($P < 0.02$ or < 0.01).

Increases can also be considered in proportion to density, and the annual (compound- interest) percentage increases are also shown in Table 1. These show the open plots (4 and 5) had the highest values, whilst Plot 1 (dense shade) had the lowest increase. These values showed significant differences between all pairs of plots at $P < 0.01$ or < 0.001.

The significant correlations (at $P < 0.05$) for the mean annual change in density on all five plots with the monthly weather variables are presented in Table 2. As yearly monitoring ceased in 1987, annual changes in density are only available for the eight years 1979-80 to 1986-87. There were 17 in two seasonal groups: positive with temperature in the previous spring; and negative with both temperature and rainfall, but positive with sunshine, in the autumn.

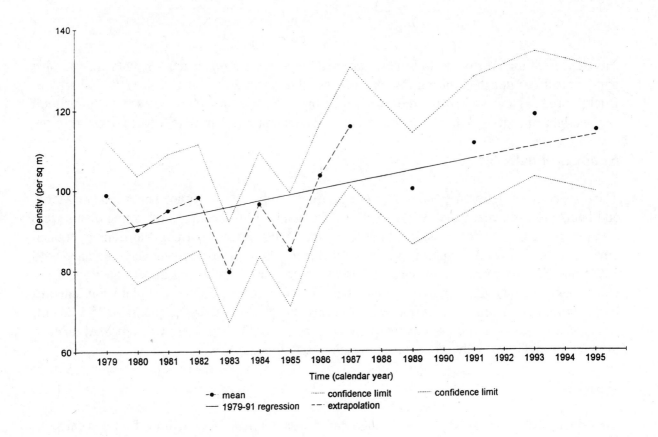

Figure 1(a). Mature bulb density (Plot 1: dense woody plants, unswiped). The confidence limit is for 95%

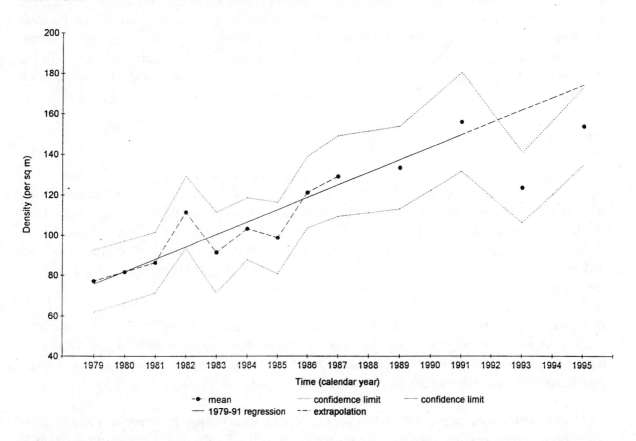

Figure 1(b). Mature bulb density (Plot 2: light woody plants, swiped). The confidence limit is for 95%

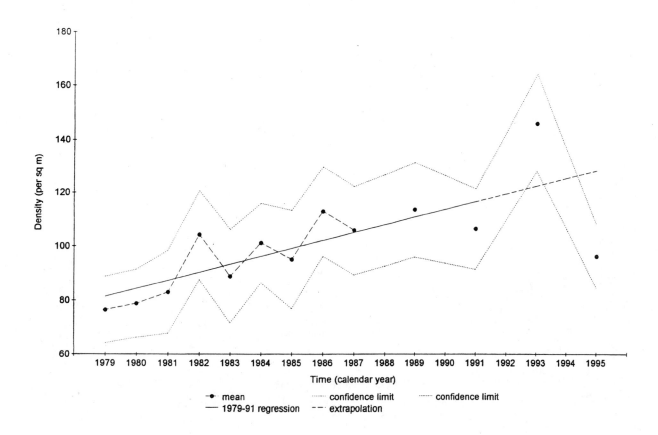

Figure 1(c). Mature bulb density (Plot 3: light woody plants, unswiped). The confidence limit is for 95%

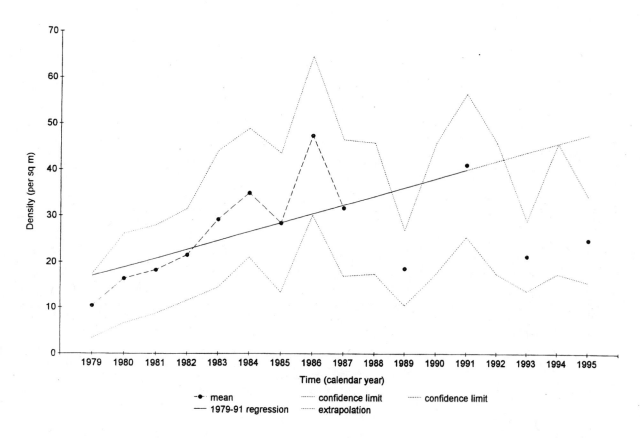

Figure 1(d). Mature bulb density (Plot 4: open, swiped). The confidence limit is for 95%.

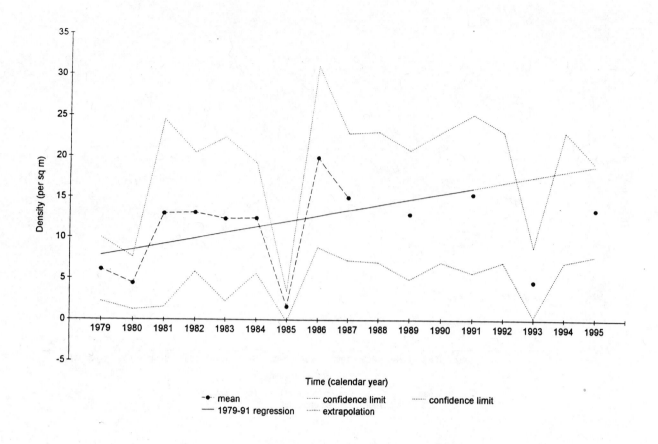

Figure 1(e). Mature bulb density (Plot 5: open, unswiped). The confidence limit for 95%.

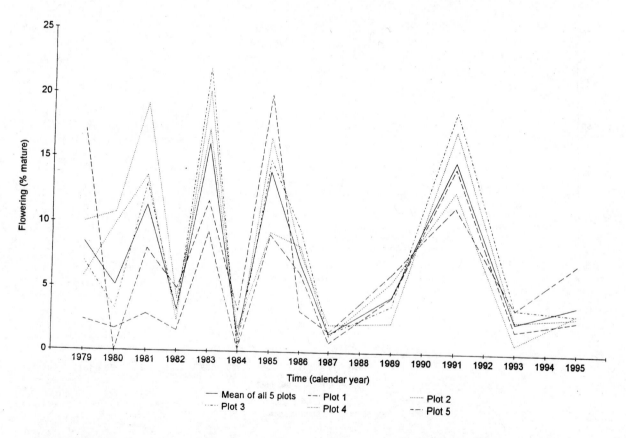

Figure 2. Flowering bulbs (% mature). (Note alternate year records from 1987).

Table 2: Statistically-significant ($P < 0.05$) correlation coefficients (r) of weather factors and changes in the density of mature daffodils (1979-87) and in the fraction of these daffodils in flower (1979-93) at West Dean Woods

(Values for which P < 0.01 are shown in bold print)

Months			Weather factor				
			Temperature			Rain	Sun
	max	min	mean	frost	soil		
Mature daffodils (mean annual increase in density)							
April		0.743		-0.758			
April-May		**0.904**		**-0.909**			
May		0.799		-0.782			
October				0.813	-0.783	0.710	
Oct-Nov		-0.833	-0.757	**0.887**	-0.778	**-0.869**	
November		-0.759	-0.762		-0.756		
Flowering daffodils (mean (%) fraction of mature)							
March							0.577
Mar-Apr						**-0.859**	0.702
April	**0.767**					**-0.745**	0.650
Mar-May				0.578		**-0.736**	
April-May						**-0.722**	
May				0.590			
July		-0.607					
February*				0.625			

* in year of flowering

Discussion on density changes

Shade

Population increase may be measured either as number per unit area (density) or in proportion to the population size. Both measures showed a considerable range. The highest increase in density was on Plot 2 (light shade, swiped), followed by Plot 3 (light shade, unswiped), with Plot 5 (open, unswiped) the lowest. Plot 5 had a dense undergrowth of partly-collapsed dead bracken fronds which over-topped many of the daffodil leaves. New bracken fronds emerge and expand as the daffodil foliage is senescing and the dead fronds acted in a similar way to the dense trees and shrubs on Plot 1. However, the two open plots are not really comparable with the others because of their very much lower densities. The highest proportional increases were in the open (Plots 4 and 5), with the least under dense shade (Plot 1).

As expected, woody-plant shade was very important for daffodils, but shade from the undergrowth also had a very strong influence. This is clear from a comparison of the pairs of swiped and unswiped plots where, in all cases, the swiped plots had higher values. Thus, the plots with dense *total* shade showed the least increases, whilst the greatest increases were on the swiped plots.

Later changes

Resulting from this conclusion, the management of the daffodil area was changed to annual swiping in mid-July (or as soon as possible afterwards) in order to have maximum effect on the bracken. Monitoring continued and the general trends of density continued but were more erratic. In 1993, three plots had densities (and 95% confidence limits) well below the extrapolated regression line - and increases in 1995. Plot 3 did the opposite, whilst Plot 1 (which could not be swiped) continued its general trend. In 1993 and 1995, most mean densities were well below the extrapolated regression lines. It seems likely that these changes were caused by the great storm of 1987 rather than from the change in management or other factors. There were some wind-blown trees on the plots with the resulting increase in microhabitats of exposed root plates and hollows. This would account for both the death of bulbs and for the increase in variation both between and within plots. By 1999 the bracken had decreased enormously and was largely replaced by grass, so that the open-plot area appeared like parkland. Monitoring is being continued and the next count, scheduled for 2000, will be interesting.

Weather

The variation within Figure 1 suggests that other factors are also operating. The most likely is the weather. With seven variables for 14 months there are 98 correlations. So, of the ten *monthly* correlations in Table 2 (which were all at $P < 0.05$) five are likely to be due to chance. The only higher-probability correlations (at $P < 0.01$) were for two-month periods. These (and most of the monthly ones) strongly suggest that spring temperature is the key weather factor controlling daffodil density. Spring is the growing season for daffodils and, clearly, it is the low temperatures (minimum temperatures and nights of frost) which are restricting their increase. In cultivation, daughter-bulb dry weight is correlated with soil temperature (Rees 1972).

Flowering

The means of flowering individuals expressed as density and as 'fraction of mature' bulbs are given in the last two lines of Table 1. On average, 8-9% of the mature bulbs flowered, except on Plot 1 (dense shade) where only 5% did so. All the plots showed large between-year fluctuations in the proportion of bulbs in flower. In most years all five plots behaved in a very similar manner (Figure 2), and the mean ranged from 1.2% (1984) to 16% (1983).

The significant correlations between flower density and weather factors for the combined plots are shown in (Table 2). Those at $P < 0.01$ were for maximum temperature and lack-of-rain in the spring and, particularly, for April of the year prior to flowering.

Discussion

The fraction flowering was least under dense shade (Plot 1) and second lowest on the open, unswiped plot with dense bracken (Plot 5). This agrees with earlier work (Salisbury 1924; Barkham 1980; Peterken 1981).

The very large between-year variation in the fraction of flowers suggests that weather, or some other random factor, is playing a more important role in controlling flowering than in governing vegetative increase.

Most of the five *monthly* correlations at $P < 0.05$ between the mean fraction-flowering and weather could be due to chance. However, all those at $P < 0.01$ strongly suggest that the weather of the spring of the year prior to flowering is very important, especially the maximum temperature in April and a lack of rain during March-May. Warm sunny weather is favourable to plant growth and, no doubt, enhances the formation of floral initials. Why spring rainfall should depress flowering the following year is less clear; possibly it is connected with the genus *Narcissus* being of Mediterranean origin. In cultivation, high temperatures followed by low temperatures are required for flower initiation and, in the field, these are satisfied by high summer and low winter soil temperatures (Rees 1972).

Density and flowering are inter-related: populations that are expanding rapidly vegetatively produce overcrowded bulbs and few flowers. Nevertheless, the new swiping management gives the best available shade treatment for both bulb density and flowering and it should be possible to inform potential visitors of the likely flowering display a year in advance.

Conclusion

After 12 years, the experimental shade treatments provided a firm basis for the sound conservation management of the wild daffodil colony in terms of both bulb density and flower production.

It was not originally intended to study the effects of weather on the daffodils but, as is often the case in monitoring, other uses of data become apparent. The literature contains many examples of how the effects of unexpected events have been able to be investigated because monitoring (which had been set up for a different purpose) was able to provide base-line data.

Acknowledgements

I thank members of the Chichester Natural History Society who did the recording; the Institute of Horticultural Research, Littlehampton, for weather data; the Sussex Wildlife Trust and West Dean Estate for permission and encouragement to carry out this work; and Richard Williamson for initially marking out and swiping the plots.

References

BARKHAM, J.P. 1980. Population dynamics of the wild daffodil (*Narcissus pseudonarcissus*) II. Changes in number of shoots and flowers, and the effects of bulb depth on growth and regeneration. *Journal of Ecology*, **68**, 635-664.

HOPKINS, B. 1999. The effect of shade and weather on daffodils *Narcissus pseudonarcissus* in West Dean Woods, West Sussex. Peterborough, *English Nature Research Reports*, No. 340.

KIRBY, K.J. & BUCKLEY, G.P. 1994. *Ecological responses to the 1987 Great Storm in the woods of south-east England.* (English Nature Science No. 23.) Peterborough, English Nature.

PETERKEN, G.F. 1981. *Woodland Conservation and Management.* London, Chapman & Hall.

POTTS, A.S. & BROWNE, T.J. 1983. The climate of Sussex. *In*: UNIVERSITY OF SUSSEX GEOGRAPHY EDITORIAL COMMITTEE, ed., *Sussex: Environment, Landscape and Society*, pp 88-108. Gloucester, Alan Sutton.

REES, A.R. 1972. *The Growth of Bulbs.* London, Academic Press.

SALISBURY, E.J. 1924. The effects of coppicing as illustrated by the woods of Hertfordshire. *Transactions of the Hertfordshire Natural History Society and Field Club*, **18**, 1-21.

STACE, C. 1991. *New Flora of the British Isles.* Cambridge, Cambridge University Press.

STEWART, J.G. 1976. *West Dean Woods Nature Reserve Management Plan.* Henfield, Unpublished ms, Sussex Trust for Wildlife Conservation Ltd.

A COMPARISON OF THE STRUCTURE AND COMPOSITION OF THE WARBURG RESERVE BETWEEN 1973 AND 1992.

K.J. Kirby & R.C. Thomas
English Nature, Northminster House, Peterborough PE1 1UA

Introduction

In this paper we give preliminary results from a comparison of the structure and composition of the Warburg Nature Reserve in south Oxfordshire in 1973 with that in 1992. This study complements work done on a similar set of plots at Wytham Woods (Kirby *et al.* 1996; Kirby & Thomas 1999). Fuller results will be published later this year as an English Nature Research Report. The Warburg Reserve, near Henley in South Oxfordshire (National Grid Reference SU715880) covers about 100 ha and was acquired by the Berkshire, Buckinghamshire and Oxfordshire Naturalists' Trust (BBONT) in 1967 (BBONT 1988).

Methods

In 1973 a 100 m grid was set up across the reserve and 10 x 10 m plots were recorded at the intersections (Dawkins & Field 1978). Each grid point formed the north-west corner of the plot. The corners of each plot were marked with underground metal markers so that they could be relocated precisely. In 1992 the plots were re-recorded. The recording on both occasions took place between April to August.

Vegetation cover was assessed by laying out a tape across the north-west to south-east diagonal of the plot. Percentage vegetation cover immediately above the tape was estimated by eye in three height bands: top or canopy cover > 2.5 m high; mid or shrub cover 0.5-2.5 m high; and ground cover < 0.5 m high. All vascular plants in the ground flora in the plot were listed but in the analysis seedlings and saplings of woody species were excluded.

Table 1. Changes in the vegetation cover across the south-east/north-west diagonal of the plot.

Layer	Mean value 1974	S.E	Mean change to 1992	S.E.
Canopy layer (>2.5 m)	72.9	3.5	-3.1	3.9
Mid-cover (0.5 - 2.5 m)	32.5	2.9	-9.6	2.8
Field and ground layer (<0.5 m)	62.4	3.6	11.1	3.2
No of plots showing cover change	>40% increase		>40% decrease	
In canopy layer	11		13	
In mid-cover	2		8	
In field and ground layer	15		3	

Results

The wood as a whole is characterised by a relatively dense canopy cover, moderate shrub cover and abundant ground layer at both dates, based on the estimates across the diagonal of the plots. The vegetation cover showed no significant change overall in the canopy cover; although there were small, but significant changes in the shrub layer and the field layer (Table 1). The mean values however hide some large changes in individual plots, with for example 25% of plots showing declines or increases of more than 40% in the canopy.

For the site as a whole the pattern of ground flora species diversity has been maintained - there is little change in the total number of species, the balance between common and scarce species, or between ancient woodland indicators and non-woodland species (Table 2). There have however been major fluctuations in the richness of individual plots.

Table 2. Comparison of species richness at plot and wood level.

	1974		1992	
Total no of species (all plots)	161		145	
No. of species recorded on only one date	37		21	
No. of species present in >10% of plots	51		53	
No. of ancient woodland indicators (no. of occurrences)	25	(228)	28	(240)
No. of other woodland species	71	(1285)	67	(1152)
No. of non-woodland species	65	(286)	50	(207)

Acknowledgements

Colyear Dawkins of the Commonwealth Forestry Institute was responsible for designing and establishing the system. We would also like to thank Nigel Phillips, Rod d'Ayala and other members of the reserve team who helped us with the recording.

References

BBONT 1988. *Warburg Reserve species handbook.* Oxford, BBONT.

DAWKINS, H.C.D. & FIELD, D.R.B. 1978. *A long-term surveillance system for British woodland vegetation.* Oxford, Commonwealth Forestry Institute (Occasional paper 1).

KIRBY, K.J. & THOMAS, R.C. 1999. Changes in the ground flora in Wytham Woods, southern England, 1974-1991, and their implications for nature conservation. Peterborough, *English Nature Research reports*, No. 320.

KIRBY, K.J. & THOMAS, R.C. & DAWKINS, H.C.1996. Monitoring of changes in tree and shrub layers in Wytham Woods (Oxfordshire), 1974-1991. *Forestry* **69**, 319-334.

LONG TERM MONITORING OF GREAT SPOTTED WOODPECKERS AND DEAD WOOD

K.W. Smith

RSPB, The Lodge, Sandy, Bedfordshire, SG19 2DL

Summary

Breeding Density of great spotted woodpecker *Dendrocopus major* has been followed at two broadleaved sites in southern England over a 16 year period. Changes in the amount of fallen dead wood and that in standing live and dead trees have been assessed over the same period. Populations of the woodpecker have been stable at one site, but increased significantly at the other. The factors limiting populations appear to differ at the two sites and over time, but include competition for nest sites with other birds and availability of standing dead trees or broken branch ends. Some lessons for other long-term studies are drawn.

Introduction

Since 1984 I have studied the breeding density, nest site selection and ecology of great spotted woodpeckers *Dendrocopus major* in two mature oak-dominated woods in Hertfordshire, southern England, (Smith 1987, 1994 and 1997). Wormley Wood, owned by the Woodland Trust since 1981, extends in total to 150ha although the study was restricted to 94ha of ancient semi-natural sessile oak *Quercus petraea*, hornbeam *Carpinus betulus* woodland. During the period of the study the management has been low intervention. Hitch Wood is a mature plantation of sessile oak, beech *Fagus sylvatica* and sweet chestnut *Castanea sativa* with some hornbeam, managed by periodic selective felling of mature trees. Fallen timber in the wood is exploited for firewood. The study was restricted to 65ha of the wood.

After intensive studies of the ecology of great spotted woodpeckers from 1984-86, long-term monitoring of breeding numbers, nest site selection and the evolution of the woodland habitat has been maintained ever since. Hence in 1999 a run of 16 years of data is available. In this paper the changes in the woodland stand composition, with emphasis on the dead wood components, and the numbers of woodpeckers are described.

Methods

Each year, intensive searches were made to locate all great spotted woodpecker nests and details of the nest tree and its diameter at breast height (dbh) recorded. Most nests were found when the adults were feeding young but the nest losses of great spotted woodpeckers are so low that well over 90% of nests would be expected to survive to this stage (Glue & Boswell 1994; Smith 1997).

At five yearly intervals (in 1986, 1991, and 1996) the overall stand structure and volumes of dead wood have been estimated for each wood. The objective has been to assess these for the whole wood so they can be related to woodpecker density and other parameters. Assessments have therefore been made at a large number of points distributed throughout each of the woods rather than collecting detailed histories in a few study plots. The methods used for the stand and dead wood assessments are summarised in Table 1. Standing trees and dead wood on the ground have been measured and assessed for condition using a four point scale similar to that adopted by Kirby *et al.* (1998). The length, diameter and condition of dead limbs on living trees were estimated from the ground. All branches at least 5cm in diameter have been included.

Table 1. Methods and sampling procedures for the stand and dead wood assessments

Variable	Measurements	Sampling regime
Standing live trees	Species, dbh, height Evidence of woodpecker holes	Within 10 m radius circle 1 pt per 4 ha in 1986 1 pt per ha in 1991 & 1996 Points on regular grid
Standing dead trees	Species, dbh, height Evidence of woodpecker holes Length, diameter, height & condition of any dead branches ≥5cm diameter	Within 10 m circle 1986, 1991 Within 25 m circle 1996 1 pt per 4 ha 1986 1 pt per ha 1991 and 1996
Dead wood on ground	Length & diameter for all ≥ 5 cm diameter Condition on 4 pt scale from fresh to highly decomposed	Within 10 m radius circle 1 pt per 4 ha in 1986 & 1991 1 pt per ha in 1996 Points on regular grid
Dead wood on live trees	Length, diameter, height & condition of any dead branches ≥5cm diameter	All live trees within 10 m circle 1 pt per 4 ha 1986 & 1991 1 pt per ha 1996

Over the course of the study it became apparent that because the spatial distribution of dead wood is very clumped a more intensive sampling regime than that originally used in 1986 was needed in order to obtain sufficient precision in the estimates. Hence the sampling strategy has evolved from one 10m radius circle per 4ha in 1986 to one per ha in 1996. Standing dead trees are in general at such low density that in 1996 they were counted within a 25m radius circle at each point representing a sampling rate of 19.6% of the area of each wood. These increases in sampling intensity will affect the precision of the overall estimates but not the mean values.

The interval of five years was selected as reasonable in terms of the effort needed and the likely time-scale of the evolution of stand structure in the woods. However each year, during the course of woodpecker work, notes are kept of management events. These notes have turned out to be extremely important particularly to assist in the interpretation of changes between each five year survey. Particularly in the managed wood a great deal can happen over the course of five years which would be extremely difficult to interpret from spot measurements.

Results

Evolution of stand composition

The overall basal areas of live trees for the major species are shown in Figure 1. For Wormley Wood there has been no significant change in the overall basal area ($F_{1,217}= 0.28$, $p=0.60$) nor that of oak ($F_{1,217}=0.34$, $p=0.56$) whereas hornbeam has increased significantly ($F_{1,217}=5.59$, $p=0.02$). For Hitch Wood the management activities have resulted in a significant decrease in the overall basal area over the period ($F_{1,150}=3.88$, $p=0.05$). However for oak and beech individually the decreases were not significant (oak, $F_{1,150}=1.38$, $p=0.24$; beech, $F_{1,150}=0.74$, $p=0.39$).

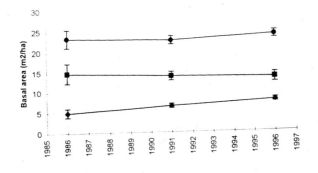

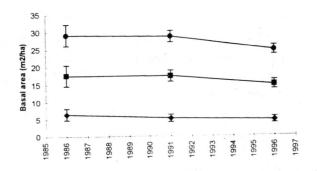

Figure 1. Basal area of living trees. Wormley Wood above; ● all trees, ■ oak, ◆hornbeam; 1986 23 points, 1991 & 1996 97 points. Hitch Wood below; ● all trees, ■ oak, ◆beech; 1986 22points, 1991 & 1996 65 points. Error bars indicate ± 1 SE

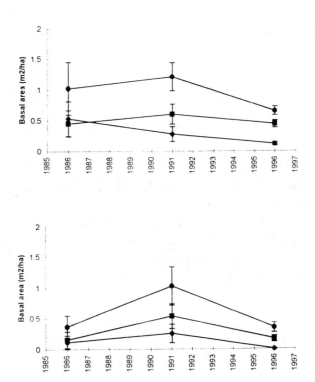

Figure 2. Basal area of standing dead trees. Wormley Wood above; ● all trees, ■ oak, ◆ birch; 1986 & 1996 23 points, 1996 97 points. Hitch Wood below; ● all trees, ■ oak, ◆ elm; 1986 & 1991 22points, 1996 65 points. Error bars indicate ± 1 SE

For standing dead trees the inadequacies of the early sampling strategy are obvious with large error bars on the results for 1986 and 1991 (Figure 2). For Wormley Wood the overall decline in basal area of dead trees just failed to attain significance ($F_{1,217}$=3.29, p=0.07) but the decline in standing dead birch was highly significant ($F_{1,217}$=5.75, p=0.02). For Hitch Wood there were no statistically significant trends in the basal area of standing dead trees, but elms had disappeared by 1996.

Dead wood volumes

The volumes of dead wood are summarised in Figure 3. In general volumes are higher in Wormley than Hitch Wood. The volume of dead wood on ground was higher in Wormley than Hitch Wood in every year except in 1996. In that year large numbers of dead limbs left after felling operations were still present in Hitch Wood but these have subsequently been removed for firewood. In both 1986 and 1991 the volumes of dead wood on living trees were similar in both woods but in Hitch Wood in 1996 this fell dramatically. The heavy thinning carried out between 1991 and 1996 probably selectively removed many oaks carrying large numbers of dead and dying limbs.

Trends in nesting great spotted woodpecker numbers

The densities of great spotted woodpeckers in the two woods from 1984 to 1999 are shown in Figure 4. In Wormley there has been an increase by a factor of two or three over the years whereas in Hitch Wood the density has stayed at a low level. In 1984 numbers were particularly low in Wormley which was thought to be a consequence of competition for nest sites from starlings *Sturnus vulgaris*. The number of starlings nesting in the wood has declined enormously since 1984 and competition is no longer a problem for the woodpeckers. In Hitch Wood nesting starling numbers were never high and there was little evidence of competition.

A wide range of nest sites have been used over the period of the study (Table 2). For both woods there has been a trend away from nesting in dead trees to living trees (Figure 5). The reasons were different in each wood. In Hitch Wood it was the loss of standing dead elms to natural decay and firewood whereas in Wormley it was loss of standing dead birch - the result of natural decay processes.

In 1994 it was reported that in Hitch Wood in particular, nest sites created by storm damage were heavily used for some years after the storms of 1987 and 1990 (Smith 1994). These sites, mainly large broken limbs on mature beech trees, are now no longer suitable and their use ceased after 1996 some six to nine years after the damaging events. This together with the loss of dead elms may now mean that nest sites are a limiting factor in Hitch Wood.

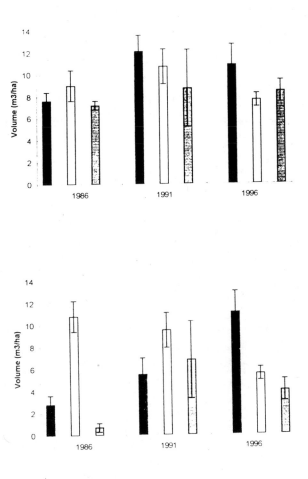

Figure 3. Estimated volumes of dead wood. Dead wood on ground (black bars), Dead wood on standing live trees (open bars), Dead wood in standing dead trees (grey bars). Wormley Wood above 25 points in 1986 & 1991, 97 points in 1996; Hitch Wood below 22 points in 1986 & 1991, 65 points in 1996.

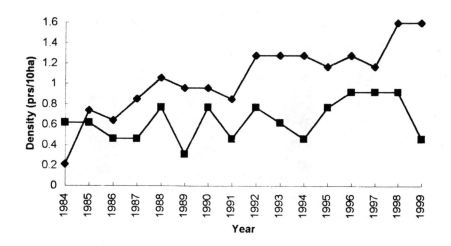

Figure 4. The density of great spotted woodpecker nests 1984-1999. Upper trace, Wormley Wood (94 ha); lower trace, Hitch Wood (65 ha).

Table 2. Nesting trees used by great spotted woodpeckers in Wormley and Hitch Woods from 1984 to 1999

Tree species	Wormley alive	Wormley dead	Hitch alive	Hitch dead
Oak *Quercus* spp.	53	12	22	2
Ash *Fraxinus excelsior*	25	0	3	0
Hornbeam *Carpinus betulus*	6	2	5	2
Beech *Fagus sylvatica*	2	0	13	2
Birch *Betula* spp.	2	49	2	1
Elm *Ulmus* spp.	0	4	0	10
Aspen *Populus tremula*	1	0	0	0
Cherry *Prunus aavilum*	0	0	1	0
Sycamore *Acer pseudoplatanus*	0	0	1	0
Larch *Larix* spp.	0	0	0	1
Scots pine *Pinus sylvestris*	0	0	0	1
TOTAL	89	67	47	19

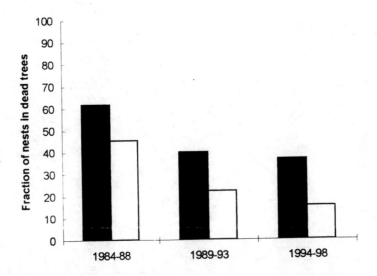

Figure 5. Fraction of nests in dead trees for each five year period. Wormley Wood (solid bars), Hitch Wood (open bars)

Discussion

Even though, from a woodland perspective, these studies have only been over a short timescale, significant and interesting trends are starting to emerge, particularly in Wormley Wood where natural changes are occurring following a long management history. For example the dead birches were almost certainly the result of seedling establishment following open canopy conditions in the 1940s and 50s. The significant increase of hornbeam, probably at the expense of oak, will be interesting to follow over the next decades. In Hitch Wood the long survival and final demise of dead elms more than 30 years after they were killed by Dutch Elm Disease is also fascinating.

The volumes of dead wood found in my studies are at the low end of the range of values reported for managed and unmanaged forests in Britain (Kirby *et al.* 1998) which is surprising given the minimum intervention in Wormley over the last few decades. Part of the differences may lie in the methods employed. Kirby *et al.* (1998) used the line-intercept method (Warren & Olsen 1964) whereas I have used full counts within circular plots. Although the line-intercept method is by far the quicker of the two it is inherently more sensitive to the exact location and orientation of the transects and therefore more open to bias.

The increase in great spotted woodpecker numbers in Wormley Wood over the 16 years was almost certainly initially a response to the removal of nest site competition from starlings but subsequently must be related to woodland habitat factors. The stability of numbers in Hitch Wood over the same period suggests strongly that management factors such as the removal of dead wood for firewood or the thinning regime are implicated. Further data from more sites would be needed to help understand the key factors. Unfortunately in long-term studies it is difficult to cover enough replicate sites to answer questions such as these.

My original question 'What determines great spotted woodpecker density?' is still difficult to answer based on these long term studies alone. Wider-scale short-term studies are now underway to help provide answers and to test ideas developed as a result of the long-term work.

However there are clear lessons which have implications for other long term studies.

i.   It is important to work out objectives, methods and sampling strategy for long-term studies at the outset so that the statistical power is sufficient to meet the objectives. This is important in any study but is particularly so in long term work where the data may not be analysed for some years. In my study I have increased the sampling intensity to improve the precision of estimates. Many long-term studies which use only a small number of small plots are likely to run into statistical problems in analysis.

ii.   In order to keep a long-term study running it must be simple and easy to do each year and it also helps if it is enjoyable. Anything too complicated or time consuming will wither and die. In my case finding great spotted woodpecker nests is easy and each year is an enjoyable challenge.

iii.   Alongside each long term study it is useful to have a series of short-term projects to look in detail at particular aspects. These can be ideal student projects in that they have clear objectives and are easy to set in the wider context.

iv.   There is the clear need for a system to register long term studies and to safeguard the data and catalogue methods. Particularly for woodland plots the timescales of long term studies are such that they are likely to outlive the researchers who started them. It will therefore be of immense value if they are documented and the data stored in such a way that the work can be repeated by future generations.

References

GLUE, D.E. & BOSWELL, T. 1994. Comparative nesting ecology of the three British breeding woodpeckers. *British Birds*, **87**, 253-268.

KIRBY, K.J., REID, C.M., THOMAS, R.C. & GOLDSMITH, F.B. 1998. Preliminary estimates of fallen dead wood and standing dead trees in managed and unmanaged forests in Britain. *Journal of Applied Ecology*, **35**, 148-155.

SMITH, K.W. 1987. Ecology of the great spotted woodpecker. *RSPB Conservation Review*, **1**, 74-77.

SMITH, K.W. 1994. The effects of the 1987 and 1990 storms on great spotted woodpecker *Dendrocopus major* numbers and nest site selection in two Hertfordshire woods. *In:* K.J. KIRBY & G.P. BUCKLEY, eds. *Ecological responses to the 1987 Great Storm in the woods of south-east England.* pp 124-133. (English Nature Science No. 23.) Peterborough, English Nature.

SMITH, K.W. 1997. Nest site selection of the great spotted woodpecker *Dendrocopus major* in two oak woods in southern England and its implications for woodland management. *Biological Conservation*, **80**, 283-288.

WARREN, W.G. & OLSEN, P.E. 1964. A line transect technique for assessing logging waste. *Forest Science*, **10**, 267-276.

MONITORING THE ENDANGERED MOTH *EUSTROMA RETICULATUM* AND ITS FOODPLANT *IMPATIENS NOLI-TANGERE* IN LAKE DISTRICT WOODLAND

Paul E. Hatcher* and **John Hooson**[#]
* Department of Agricultural Botany, School of Plant Sciences, The University of Reading, 2 Earley Gate, Whiteknights, Reading, RG6 6AU, UK.
# The National Trust, North West Region, The Hollens, Grasmere, Ambleside, Cumbria, LA22 9QZ, UK.

Summary

The netted carpet moth (*Eustroma reticulatum*) is confined in the UK to the Lake District (apart from a couple of specimens recorded from mid-Wales). Its larvae feed only on touch-me-not balsam (*Impatiens noli-tangere*), itself a nationally scarce plant that mainly occurs in woodland. In 1990 a full survey of Lake District sites was carried out (following surveys in 1955 and 1980) and since 1993 areas alongside Derwent and Coniston Water have been surveyed annually. These surveys suggest that the size of colonies can vary by an order of magnitude between years and that individual colonies may exist for only a couple of years. The smaller the foodplant colony the less likely it is to contain moths and the more likely it is to go extinct. Management may not prevent extinction but should seek to maximise colony formation.

Introduction

The moth and its foodplant

The netted carpet *Eustroma reticulatum* (Lepidoptera: Geometridae) is confined to touch-me-not balsam *Impatiens noli-tangere* as a foodplant. The moth is univoltine, flying between late June and mid-August and can be observed at dusk near colonies of foodplants. Eggs are laid singly on the leaves and soon hatch. The pale yellow/green caterpillars feed preferentially on flowers and seed pods but also on leaves and are fully grown by mid-September. By October the caterpillar has pupated in the soil where it overwinters.

Impatiens noli-tangere (Balsaminaceae) is an annual that occurs in damp open woodland (chiefly W9 ash - rowan - dog's mercury or W7 alder - ash - yellow pimpernel woods) where it favours streamsides, moist shady roadsides, sites of disturbance (e.g. logging operations, windthrown trees), and occasionally gardens. It generally grows on silty soils with a high organic content and flowers from early July through August, although in the shade reproduction is mainly by cleistogomous flowers. It does not have a persistent seed bank, and thus is dependent on setting seed each year (Markov 1991).

Impatiens noli-tangere is a nationally scarce species. In the UK it is probably native (occurring in 16 10km squares) only in the Lake District, around Dolgellau and in a small area on the Montgomery - Shropshire border, where it was found new to Britain in 1632 (Coombe 1956; Stewart, Pearman & Preston 1994). It also occurs in 78 10km squares as an introduction (Stewart, Pearman & Preston 1994), although many of these records may refer to the introduced orange balsam *I. capensis* or small balsam *I. parviflora*.

History

The history of the moth in the Lake District has been well documented by Birkett (1951), Heath (1959, 1983) and Hatcher and Alexander (1994). *Eustroma reticulatum* was first discovered in the UK by T. H. Allis in August 1856 in the Claife woodland on the north-west shore of Lake Windermere and reported by Doubleday (1861). It was not re-found until 1876, when it was discovered on its hitherto unknown foodplant. Up to six sites for the moth were known around Lake Windermere until the turn of the century, when it was thought to have become extinct. A later site existed until 1923, after which it was thought that the moth had become extinct again. During this time, sites seldom existed for many years, with reports of sites being 'destroyed' or plants 'removed'. However, the sporadic recording of the moth during this period is undoubtedly due to the small number of sites examined by entomologists, who tended to visit known sites in preference to searching for new ones.

Although foodplant sites were searched in 1940, the moth was not rediscovered until 1945 (Birkett 1951). The first systematic survey of all known Lake District sites was carried out by Heath (1959) in 1955, and repeated in 1980 (Heath 1983). Eight of the sites where the moth occurred in the 1980 survey were resurveyed in 1989, and larvae were found in only half the sites. This prompted a survey of all known sites in 1990 (Hatcher 1991; Hatcher & Alexander 1994). After a preliminary re-survey in 1993, sites along Coniston and Derwent Water were surveyed in 1994, 1995 and 1996 under English Nature's Species Recovery Programme. In 1995 the moth was included in the short list of species of greatest conservation concern in the UK Biodiversity Action Plan (one of only three moths included) and in 1997 and 1998 the Coniston and Derwent Water sites continued to be surveyed under this scheme. We concentrate here on these sites surveyed annually, although other sites were also resurveyed between 1990 and 1998.

The only other records for *E. reticulatum* in the UK are from mid-Wales, where occasional moths have been recorded since 1886. Hatcher and Alexander (1994) provide a discussion of early Welsh surveys, while Howe and Fowles (1998) and Hammett and Hull (1999) describe later surveys.

Methods

All sites surveyed by Heath (1959, 1983) and sites subsequently noted by local naturalists were surveyed between 30 August and 9 September 1990 (Hatcher & Alexander 1994). At this time of year the larvae are usually reaching their final instar and are easy to spot on the undersides of leaves and flower pods.

At each site the number of foodplants was estimated and sample plants were examined for larvae. By counting the number of larvae on an estimated proportion of the plant population (if large) or on all plants, an estimate of the population size of larvae was made. In all cases, all foodplants were examined if no larvae were found. Where possible, sketch maps were made of the colonies, photographs taken and sometimes pegs were driven into the ground to mark the position of small colonies.

The 1993 to 1998 surveys used the above methods but concentrated on sites on the east shore of Coniston Water and the east and west shores of Derwent Water. These areas have a number of sites within woodland mainly owned by the National Trust and thus were amenable to sampling.

In addition, the areas around the Coniston and Derwent Water areas were searched for further colonies.

The basic unit of description in the 1955, 1980 and 1990 surveys was the 'site' - a more or less homogeneous area containing the moth or foodplant, following Heath (1959, 1983). There can be one or more 'site' per 'area' (e.g. east Derwent Water). Since 1990 the 'colony' (a distinct clump of plants separated from other colonies by up to 100m) has been used as the basic unit of measurement. Thus comparisons at three levels: 'area', 'site' and 'colony' are possible.

Results

The present distribution of the moth and foodplant is given in Figure 1. Colonies are concentrated in eight main areas: Coniston Water ; east Lake Windermere; west Lake Windermere; Ambleside; Bridgend; east Derwent Water; west Derwent Water and Muncaster.

The 1990 survey

In 1955, 16 sites contained both moth and foodplant, and three contained only the foodplant (Table 1). In 1980 of the 18 sites re-sampled, 14 contained moths, three had foodplants only and one had neither. However, nine new moth sites and three new foodplant sites were also sampled. In 1990, only 11 of these sites still contained moth colonies, nine contained foodplants only and eleven sites had neither. Three new moth sites and three new foodplant sites were also found. Overall, this represented a 56% decline in moth sites, falling below 1955 levels (Hatcher & Alexander 1994). Over 80% of sites were directly associated with woodland, and all but 15% were under some form of shade.

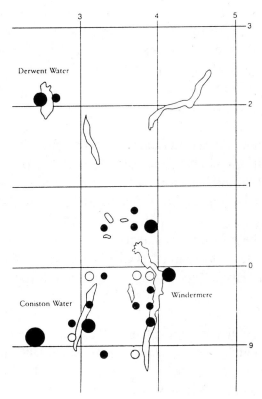

Figure 1. Distribution by tetrad (2km x 2km square) since 1990 of the netted carpet and foodplant in the Lake District. Filled circles indicate moths present, the larger the circle the greater the population, open circles indicate foodplant only present.

Table 1. Summary of *Eustroma reticulatum* surveys in the Lake District (from Hatcher & Alexander 1994).

Date of survey	Number of sites		Sites without moths (%)
	With moths	Foodplants only	
1955	16	3	16
1980	23	6	21
1990	14	12	46

Table 2. 95% confidence intervals (CE) for plant and moth populations in Coniston and Derwent Water areas, calculated from 1990 - 1997 data, compared with recorded values for 1998.

	Plant		Moth	
	95% CE	1998	95% CE	1998
Coniston	1360 - 2606	341	123 - 565	92
Derwent	1882 - 3496	134	194 - 322	62

Fluctuations at area and site level since 1990

Coniston Water This area contains eight to ten distinct sites and numerous colonies scattered over a 3km length of woodland.

Between 1990 and 1993, there was a 125% increase, and between 1993 and 1994 a 25% increase in plant numbers and a 575% increase in caterpillars between 1990 and 1994 (Figure 2a). Although some new sites were found between 1991 and 1993, this increase was due mainly to population increases in known sites (Figure 2b). Management of these sites was carried out between 1991 and 1993, which helped increase population sizes.

There was a significant decrease in foodplant (65%) and moth (84%) numbers between 1994 and 1995. In 1998 foodplant numbers were only 21% of the previous, poor, year, and 11% of the best year (1994) and moth numbers were 43% of the previous year (Figure 2a) - outside the 95% confidence intervals for populations calculated from 1990 - 1997 surveys (Table 2).

Derwent Water The two areas east Derwent Water and west Derwent Water are combined in this analysis. East Derwent Water has one site in shore-side woodland (Figure 3) containing several colonies, while west Derwent Water contains one main site in shore-side woodland (Figure 4) with two smaller sites nearby.

Derwent Water sites show a similar trend to Coniston Water sites, with an increase of 45% in foodplant numbers and 75% in moth numbers between 1990 and 1994 (Figure 5). Between 1990 and 1994 no new sites were discovered, but the removal of dead elms opened up the canopy for the plant to spread, and coppicing of alders in the north-east of the area also allowed colonies of balsam to spread. The decrease in plant numbers in 1995 was not as pronounced as at Coniston Water, and was not accompanied by a decrease in moth numbers (Figure 5). The following year the plant numbers had increased by 137%. In 1997 and 1998 there has been a decline in both plant and moth numbers, even greater than at Coniston Water. In 1998 only 4% of the foodplants recorded in 1997 occurred and only 24% of the caterpillars; again these values are outside the 95% confidence intervals for population size (Table 2).

a)

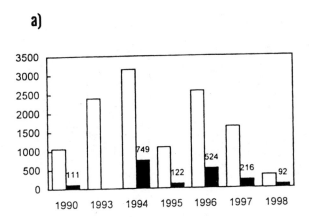

b)

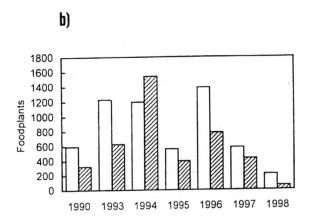

Figure 2. (a) Total number of foodplants and estimated caterpillar populations in the Coniston area 1990-1998. Open bars indicate numbers of foodplants, closed bars indicate estimated number of caterpillars (with numbers given above). Note that moth populations were not estimated in 1993. (b) Numbers of plants recorded from the two major sites within the Coniston area. Open bars: Site *a*; hatched bars: Site *b*.

The moth density at Derwent Water often shows an inverse relationship with plant numbers: for example, good years for the plant had low moth density (1994: 9 larvae per 100 plants; 1996, 8; 1997, 9) while poor years for the foodplant had a high density (1995, 17; 1998, 46).

Colony survival

Between 1993 and 1997, the size of colonies in the Derwent and Coniston Water areas varied between 1 and over 1000 plants, but most had less than 150 plants (Figure 6). There was considerable year-to-year variation in colony size: this is illustrated for the east Derwent Water (Figure 3) and west Derwent Water colonies (Figure 4) where the small area occupied and abundance of unchanging topographical reference points made the mapping of populations on to sketch maps quick and relatively accurate. Also, one can be certain that no colonies were overlooked within this area. These figures also illustrate the large reduction in colony sizes between 1997 and 1998.

75

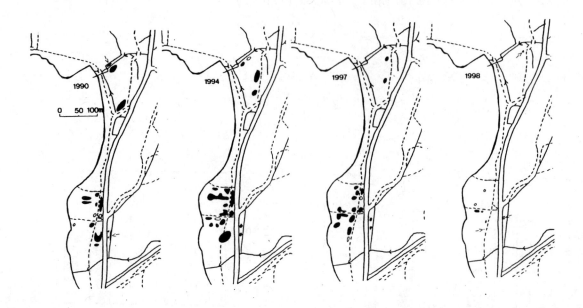

Figure 3. Sketch maps of east Derwent Water colonies 1990 - 1998. Filled areas indicate caterpillars present, open areas indicate foodplants only. Arrows indicate very small colonies.

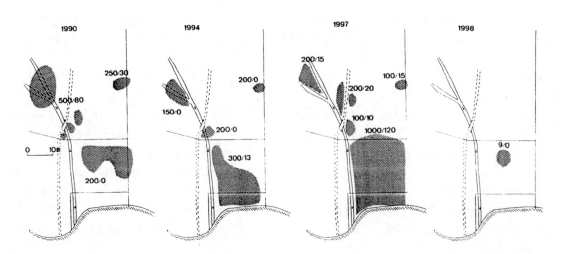

Figure 4. Sketch maps of west Derwent Water, Kitchen Bay colonies 1990 - 1998. Shading indicates area covered by foodplants, estimated numbers of foodplants/number of caterpillars given in bold for each colony.

Loss of moths: Between 1993 and 1997, 89 colony records (a colony record is one colony recorded in one year; thus if a colony existed between 1993 and 1997 it would have 5 colony records) were without moths (42%). On 24 occasions a colony had lost the moth compared to 56 occasions when a colony had retained moths from one year to the next.

A greater proportion of smaller colonies were without moths (Figure 6) than larger colonies and colonies which lost the moth had smaller numbers of moths (17.4 ± 5.09, $t = 2.18$, $P = 0.034$) and foodplants (116 ± 31.6, $t = 2.55$, $P = 0.013$) the year before compared with colonies that retained them (mean no $\pm$ SE, moths 44.1 ± 11.1, foodplants 230 ± 31.7). During 1993 - 1997, 11 colonies regained moths after not having them the year before.

Loss of foodplant Twenty-nine colonies were recorded in the Derwent and Coniston areas in the 1990 survey, and 32 new ones were found between 1991 and 1997. During this period 27 colonies became extinct. Although moths can reappear at a colony they were not recorded at the previous year, this has not been observed for the foodplant. Plant colonies which became extinct had significantly fewer plants the year before (33 ± 6.7, $t = 5.66$, $P < 0.001$) than colonies which remained (142 ± 18.2). Between 1997 and 1998 the foodplant became extinct from a further 15 colonies, including for the first time colonies which contained over 100 plants the previous year (215, 200 and 420 plants, respectively).

It is difficult to measure colony longevity but some indications can be gained from colonies that have been formed along Coniston Water from road-sweepings dumped in car-parks or by the side of the road. In one case foodplants (less than 10) were present for seven years; in another plants have been present for at least 10 years, but reduced from 90 to 10; and in another plants declined from 300 to zero over six years.

Conclusions - surveys and management implications

The results presented here emphasise what can be achieved with a very limited budget and very limited time. Each yearly survey between 1993 and 1998 was carried out in less than one week, and yet they are starting to reveal patterns in colony survival and change that would be lost in larger-scale once-every-ten-years surveys. We hope one of these larger surveys will be possible in 2000 for comparison with the previous surveys.

The results from surveys and associated management work suggest that one aim should be to ensure that sufficient large foodplant colonies (i.e. over 200 plants) exist within each area. Smaller colonies are much more likely to become extinct, to lose the moth, or not to be colonised by the moth. Small colonies are also much harder to manage to increase their foodplant numbers. However, each area is a dynamic system of colonies, each of which should not be expected to persist for many years without appropriate management. This will not be a problem as long as sufficient new colonies can be established either naturally or artificially.

We hope to continue the yearly surveys into the next millennium: the next couple of years will be particularly interesting as we see if the moth and foodplant can recover from their very low numbers in 1998.

77

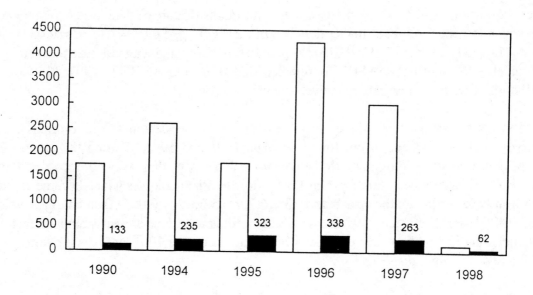

Figure 5. Total number of foodplants and estimated caterpillar populations in the east and west Derwent Water areas. Open bars indicate numbers of foodplants, black bars indicate estimated caterpillar populations (with numbers shown above bars).

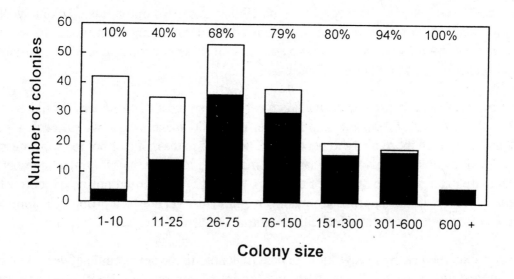

Figure 6. Distribution of colony sizes at Derwent and Coniston Waters 1993 - 1997. Open bars indicate all colonies, filled bars indicate colonies with moths (percentages given above bars). Colonies tallied for each year of existence. Chi-squared statistic for difference between colonies with moths and those without = 71.36, $P<0.0001$, 5 d.f.

Postscript

Subsequent to the first preparation of this paper, the 1999 Lake District survey has taken place. At Derwent Water 98 plants were found with an estimated larval population of five, while at Coniston Water 1600 plants were found with an estimated larval population of 102. It is encouraging that the number of plants at Coniston Water has increased to a level last recorded in 1997 (although the moth population has improved little since 1998), but the Derwent Water totals are worse even than 1998, and there must now be considerable concern for the survival of the moth in this area.

Acknowledgements

The 1990 survey was funded by the World Wide Fund for Nature and the 1993-1999 surveys by The National Trust and English Nature, to whom we are grateful. We wish to thank all the landowners who have allowed us access to sites and the members of the Netted Carpet Moth Biodiversity Action Plan Steering Group and especially Butterfly Conservation who organised the steering group. Finally, we wish to thank everyone who has sent us records for the moth and foodplant in the Lake District.

References

BIRKETT, N.L. 1951. A short history of *Eustroma reticulata* Schiff. *Transactions of the Raven Entomological and Natural History Society,* **1951,** 37-39.

COOMBE, D.E. 1956. Notes on some British plants seen in Austria. *Veröffentlichungen des Geobotanischen Instituts, Eidgenössiche technische Hochschule Rübel in Zürich,* **35,** 128-137.

DOUBLEDAY, H. 1861. Occurrence of *Cidaria reticulata* in the Lake District. *Zoologist,* **19,** 7361.

HAMMETT, M.J., & HULL, M. 1999. *A survey for the netted carpet moth* Eustroma reticulata *at sites in Merionethshire (1998).* Bangor, Countryside Council for Wales (Contract Science Report, No. 329).

HATCHER, P.E. 1991. *The netted carpet moth: an assessment of the current status and management requirements in the Lake District.* London, Report to the Joint Committee for the Conservation of British Insects.

HATCHER, P.E., & ALEXANDER, K.N.A. 1994. The status and conservation of the netted carpet *Eustroma reticulatum* (Denis & Schiffermüller, 1775) (Lepidoptera: Geometridae), a threatened moth species in Britain. *Biological Conservation,* **67,** 41-47.

HEATH, J. 1959. The autecology of *Eustroma reticulata* Schiff. (Lepidoptera: Geometridae) in the Lake District with notes on its protection. *Journal of the Society of British Entomology,* **6,** 45-51.

HEATH, J. 1983. The insects of the yellow balsam, *Impatiens noli-tangere. Proceedings and Transactions of the British Entomological and Natural History Society,* **16,** 125-131.

HOWE, M.A., & FOWLES, A.P. 1998. *The status of the netted carpet moth* Eustroma reticulata *in Merionethshire*. Bangor, Countryside Council for Wales (Natural Science Report, No. 98/5/1).

MARKOV, M.V. 1991. Population biology of touch-me-not *Impatiens noli-tangere* L. *Soviet Journal of Ecology*, **22**, 12-20.

STEWART, A., PEARMAN, D.A., & PRESTON, C.D. 1994. *Scarce Plants in Britain*. Peterborough, Joint Nature Conservation Committee.

RESTORATION OF WOOD-PASTURE IN BURNHAM BEECHES: SOME PRELIMINARY RESULTS FOR PLANTS AND GROUND RUNNING INVERTEBRATES

Helen J. Read
Corporation of London, Hawthorn Lane, Farnham Common, Bucks. SL2 3TE

Introduction

An area of 6 ha of neglected wood-pasture within Burnham Beeches is being restored to what we think was more like its state 100 years ago. Prior to 1989 it consisted of veteran pollards, both beech *Fagus sylvatica* and oak *Quercus* spp., surrounded by dense secondary woodland of holly *Ilex aquifolium* and birch *Betula* spp. Between 1989 and 1994 progressive clearance of trees and shrubs left an open woodland of veteran trees and some young beech and oak pollards. Seasonal grazing with ponies, cattle, sheep and pigs, was re-introduced in 1992. The area has been flailed to control birch regeneration and sprayed/rolled to control bracken. A monitoring programme was set up between 1990 and 1992 to record changes in the vegetation and ground-running invertebrates.

Methods

Vegetation is assessed in three 30x30m squares marked out in the restored area and one in unmanaged woodland. In each of these 25 0.5x0.5m quadrats are located at random each June. Percentage cover of plant species in the quadrats is estimated and a list made of all species. Results are presented from one of the squares in the restored area and the control (Table 1).

Invertebrates are assessed using a grid of 10 pitfall traps (plastic cups) set in the area being restored with another in the control area. Each trap contains 5ml of 4% formalin with detergent added and has a wooden roof. Traps are emptied every two weeks from April to October. All invertebrates are identified to species. Results are presented for spiders only in Table 2.

Results

The restored wood-pasture was cleared in two phases, winter 1991/1992 and March 1994. The number of trees/shrubs in the vegetation plot was reduced from 19 to 6 and shrub layer holly cover from 20% to a single small bush. Beech and oak were retained, both as old pollards and young maidens. Between 1992 and 1999 the control plot showed an increase in canopy cover from 60% to 80% and the number of trees/shrubs rose from 57 to over 80 (mostly birch and holly saplings).

Reduction in tree cover resulted in an increase in field layer species, especially grasses. Common bent *Agrostis capillaris* showed the biggest increase in cover, but several years after clearing species such as heath grass *Danthonia decumbens* and fine-leaved sheep's fescue *Festuca tenuifolia* were recorded. Bracken increased after clearance but due to the control methods carried out this is not obvious. The amount of bramble did not increase significantly in this sample, although it did elsewhere. Within the control plot there is variation from year to year, but plant species involved were mainly ruderals such as greater plantain *Plantago major* and annual meadow-grass *Poa annua* rather than acid grassland or heathland species.

The number of wolf spider (Lycosidae) individuals increased following clearance, while the number of money spiders (Linyphiidae) decreased. Some species showed very pronounced patterns, *Alopecosa pulverulenta* and *Pachygnatha degeeri* both increased during restoration while *Lepthyphantes flavipes* and *Diplocephalus picinus* decreased.

Conclusions

Substantial changes have ben recorded in the vegetation and ground running invertebrates during restoration of wood-pasture from secondary woodland, reflecting the more open conditions compared to the control woodland where the canopy cover increased over time.

Table 1. Selected results for Vegetation

Restored wood-pasture	1990	1991	1992	1993	1994	1995	1996	1997	1998	1999
No. vascular plant spp.	14	10	13	16	16	22	16	23	22	23
No. grass spp.	4	3	3	4	4	8	7	9	8	8
Mean % Cover of:										
Bare ground	65.40	58.00	30.90	54.60	68.60	61.10	39.70	35.10	14.80	12
Rubus fruticosus	0.00	1.00	0.60	0.00	0.00	0.10	1.08	1.10	2.70	0.4
Pteridium aquilinum	5.40	16.40	15.80	7.10	0.00	0.70	2.84	0.40	13.90	14.1
Ilex aquifolia	14.08	15.76	19.20	6.32	2.96	0.88	5.44	0.28	3.68	1.24
Deschampsia flexuosa	15.28	23.32	20.20	9.52	5.00	25.20	14.80	15.24	14.20	22.16
Agrostis capillaris	0.00	0.00	8.80	7.40	7.40	20.20	16.40	20.96	34.48	39.72
Control Woodland										
No. vascular plant spp.	NR	NR	9	9	18	11	8	11	14	15
No. grass spp.			1	1	3	1	1	2	3	4
Mean % cover of:										
Bare ground			86.80	64.00	77.80	83.90	47.00	50.00	62.50	46.30
Rubus fruticosus			0.40	0.00	0.60	0.00	0.00	0.00	0.10	0.40
Pteridium aquilinum			18.70	18.70	3.40	3.20	26.50	25.80	9.60	14.50
Ilex aquifolium			2.80	5.68	9.40	8.36	9.72	12.44	10.84	15.35
Deschampsia flexuosa			2.44	2.80	2.48	2.04	5.64	1.60	6.04	4.58
Agrostis capillaris			0.00	0.00	0.20	0.00	0.00	P	P	0.58

(P = present, NR = not recorded)

Table 2. Selected results for spiders

Restored wood-pasture	1990	1991	1992	1993	1994	1995	1996	1997	1998
No. of weeks	22	23	22	22	22	22	22	22	22
Spider individuals	1192	1703	816	879	1116	2162	2465	1185	2107
Spider species	50	40	28	41	45	45	37	41	41
Lycosidae individuals	689	1353	691	665	977	1871	2274	1003	1826
Linyphiidae individuals	433	296	107	185	82	143	69	60	120
Pardosa lugubris	618	1300	599	535	518	945	1009	89	47
Alopecosa pulverulenta	11	0	31	35	129	231	551	362	800
Pachygnatha degeeri	0	0	0	1	16	39	31	21	125
Diplocephalus picinus	185	94	4	11	3	35	13	3	0
Lepthyphantes flavipes	59	84	32	10	18	23	16	2	3
Control woodland									
No. of weeks			16	22	22	22	20	22	22
Spider individuals			94	574	518	1549	677	416	533
Spider species			14	24	30	34	28	20	26
Lycosidae individuals			48	376	295	1027	327	254	340
Linyphiidae individuals			40	181	215	582	325	154	182
Pardosa lugubris			45	126	262	980	299	146	299
Alopecosa pulverulenta			0	3	2	16	5	84	7
Pachygnatha degeeri			0	0	1	1	0	0	0
Diplocephalus picinus			0	11	12	133	50	11	50
Lepthyphantes flavipes			22	49	29	88	21	41	18

NR = not recorded

SURFACE ACTIVE ARTHROPODS IN A CHRONOSEQUENCE OF SCOTS PINE (*Pinus sylvestris*) IN THE NEW FOREST

Tilly Collins
School of Biological Sciences, University of Sussex BN1 9QG, current address Imperial College, Silwood Park SL5 7PY.

Introduction

In the New Forest, Hampshire, commercial conifer monoculture forestry, commonly held to be a species-impoverished environment (Thornber, Legg & Malcolm 1993), is mixed in an intimate mosaic with one of England's ancient semi-natural forests. The physical changes associated with crop development are accompanied by a changing arthropod fauna (Day & Carthy 1988). In this study chronosequence 'snapshot' data contributes to assessing diversity over a rotation, as long term monitoring data is not available. The arthropod order and carabid species diversity measures are predicted to rise with crop age.

Entomological data is now commonly used as an indicator of the biological status and conservation value of a site (Refseth 1980, Butterfield *et al.* 1995). Carabid beetles have been selected because they are well known taxonomically, widely distributed, easily trapped and are largely polyphagous, thus reflecting much ecological information from their communities. Sites were selected using the Ecological Site Classification (ESC) method (Hodge 1995) to represent a chronosequence within a patch clearfell management system(Table 1).

Table 1: Crop characteristics of the sites. DBH (± SE) is the diameter at breast height of a tree.

	Site 1	Site 2	Site 3
Crop species	*P. Sylvestris*	*P. Sylvestris*	*P. Sylvestris*
Planting date	1970	1947	1930
Age	26	49	66
Forestry class	Thicket	Mid-rotation	Mature
Stocking level (stem/ha)	6600	850	250
Mean DBH	8.65 (± 0.950	27.63 (± 0.79)	42.97 (± 1.64)

Materials & Methods

Pitfall trapping was chosen to sample surface-active arthropods. Plastic 200ml cups were perforated seven cm from the base, half filled with water and 0.1% Decon 90 and covered with green tin sheets raised four cm from the ground. Two parallel rows of eight traps placed three metres apart in the centre of the block were set on 7/8/1996. Seven weekly collections were made and the catch stored in 90% Industrial Methylated Spirit.

Identification of the whole catch, excluding acari and pseudoscorpionids, was made to Class (Diplopoda, Chilopoda & Collembola) or Order level and carabid beetles were taken to species level (Tilling 1987; Forsythe 1987; Lindroth 1974).

In addition to the two simple measures of diversity, Species number (S) and number of individuals (N), the Shannon Index (H) and the Shannon measure of evenness (J) (Magurran 1988) were calculated for each site at both levels of data resolution. The Brillouin index (Hb), which accounts for non-randomness in a sample due to the activity budgets of different species leading to varying probabilities of capture, was also calculated for carabid species level data.

Results

Order level data showed a pattern of rise in diversity with age between sites 1 and 2 in both N and H. This pattern altered abruptly at site three (Table 2). Ants rise from 3% of the catch at site one to 84% of the catch at this site, which reduced H and J to less than half that of the younger sites. If the ant data are discounted, these measures rise with plantation age.

Table 2: Measures of diversity for the whole catch at Order level (ants discounted)

Total Catch	Site 1	Site 2	Site 3
Number of individuals	1348 (1307)	1883 (1618)	11,904 (1866)
Number of orders	13 (12)	12 (11)	12 (11)
Shannon index (H)	1.788 (1.716)	1.822 (1.816)	0.698 (1.958)
Shannon evenness (J)	0.697 (0.690)	0.733 (0.757)	0.281 (0.816)

Carabid species level data indicated a rise in H, Hb and J with plantation age (Table 3). The number of individuals trapped at the mid-aged site was double that of either the thicket or mature sites. S was significantly different between sites (ANOVA, F=22.7, P,0.0001, df 63). All three sites were dominated by *Abax parallelopipedus,* which represented over 70% of the catch at all sites. In total 10 carabid species were found.

Table 3: Measures of diversity for the three sites at Carabid species level

Carabidae	Site 1	Site 2	Site 3
Number of individuals	180	386	187
Number of species	4	7	6
Brillouin index (Hb)	0.43	0.71	0.75
Shannon index (H)	0.464	0.734	0.789
Shannon evenness (J)	0.335	0.377	0.44

Discussion

Suggestions that coniferous stands (Day, Marshall & Heaney 1993; Evans 1987) possess a greater arthropod diversity than we think are not greatly supported here. However, the relatively poor carabid fauna collected across the chronosequence may be partly explained by the short time allowed for trapping and the limited number of sites available in the chronosequence. A recently clearfelled site would have provided a dramatically different habitat type. The unexpectedly impoverished carabid S and N trapped at site three may well be due to competition or interference from wood ants (Hawes, pers. Comm.) which were particularly abundant at this site.

Acknowledgements

Thanks are due to: Forest Research BRP, Cathy Hawes, Martin Jukes, Simon Leather and Alan Stewart.

References

BUTTERFIELD, J., LUFF, M.L. BAINES, M. & D. 1995. Carabid beetle communities as indicators of conservation potential in upland forests. *Forest Ecology & Management,* **79,** 63-77.

DAY, K.R. & CARTHY, J. 1988. Changes in Carabid Beetles Communities Accompanying a Rotation of Sitka Spruce. *Agriculture, Ecosystems and Environment,* **24,** 407-415.

DAY, K.R., MARSHALL, S. & HEANEY, C. 1993. Associations between forest type and invertebrates; ground beetle community patterns in a natural oakwood and juxtaposed conifer plantations, *Forestry,* **66,** 37-50.

EVANS, H.F. 1987. Sitka Spruce insects: past present and future. *Proceedings of the Royal Society of Edinburgh,* **93B,** 157-167.

FORSYTHE, T.G. 1987. *Common Ground Beetles.* Naturalists' Handbook 8. Slough, Richmond Publishing Co. Ltd.

HODGE, S. 1995. *Biodiversity Research Programme, April 1995 to March 1998.* Edinburgh, Forestry Authority Research Division.

LINDROTH, C.H. 1974. *Handbooks for the identification of British Insects, Vol. IV (2): Coleoptera, Carabidae.* London, Royal Entomological Society of London.

MAGURRAN, A.E. 1988. *Ecological Diversity and its Measurement.* London, Chapman & Hall.

REFSETH, D. 1980. Ecological analyses of carabid communities - Potential use in biological classification for conservation. *Biological Conservation,* **17,** 131-141.

THORNBER, K.A., LEGG, C.J. & MALCOLM, D.C. 1993. *The influence of stand manipulation on the biodiversity of managed forests: a review.* Edinburgh, Institute of Ecology and Resource Management.

TILLING, M.S. 1987. A key to the major groups of terrestrial invertebrates. *Field Studies,* **6,** 695-766.

TWO DECADES OF DATA ON OAK DEFOLIATION IN A WORCESTERSHIRE WOODLAND NNR

David Harding

School of Applied Sciences, University of Wolverhampton WV1 1SB; current address Cliff Cottage, Woodhouse Lane, Boningale, Albrighton, Wolverhampton, WV7 3BZ

Summary

Monitoring of half-a-dozen individual oaks in Chaddesley Woods, Worcestershire, has revealed marked variations within and between the 1980s and the 1990s in the amount of defoliation by caterpillars. Annual averages always exceeded 30% in the 1980s, with peaks in 1980 and 1990, whereas they have been less than 10% since 1991. This might be associated with climate change, or alternatively with an underlying 10-year cycle of caterpillar abundance.

Introduction

Chaddesley Woods NNR, near Kidderminster, provided a classic example of defoliation during May, 1980: virtually all of the oaks and other broad-leaved trees and shrubs were stripped of their leaves by caterpillars (Packham *et al*. 1992). A long-term project was begun in 1982, studying the state of the canopy of individual oak trees. Initially, the object was to follow up the work of Satchell (1962) on the extent of defoliation relative to the flushing sequence. Subsequently this was supplemented by monitoring of caterpillar abundance, for comparison with other long-term studies (Varley, Gradwell & Hassell 1973). Such data are not only of value in their own right, but inevitably raise questions as to the causes of fluctuations (Berryman 1996), including the possible role of climate change, with potential knock-on effects elsewhere in the food-web, eg titmice (Visser *et al*. 1998).

Materials and methods

Chaddesley Woods (SO 914736) represent an ancient woodland fragment of the medieval Forest of Feckenham, with *c*50 ha mainly under oaks (*Quercus robur* on Keuper Marl, *Q. petraea* on more sandy soils), with an understorey of neglected hazel coppice (*Corylus avellana*). It was the first NNR to be designated under the Nature Conservancy Council in 1973, being subsequently managed by English Nature and, more recently, by Worcestershire Wildlife Trust.

A clearing, *c*100 x 60m, was created in compartment 2 in the late 1970s to encourage regeneration of oaks, one half being protected as a rabbit-proof exclosure. Sixteen oaks *Q. robur* within and around this clearing were selected for study, of which seven have provided detailed records. Their average dbh is 40cm, and their height *c*15m. They probably originated from coppice stools during the nineteenth century.

During 1982-84, and every year from 1987 to date, observations have been made of each tree at approximately weekly intervals between mid to late April (to record the spring flush) and at least the middle of June, when regrowth foliage ('midsummer flush') first appears. On each occasion, subjective estimates were made of the amount of development of the foliage of each tree. At the start of the season this was expressed in terms of the degree of bursting of the buds, thus establishing the flushing sequence for the seven trees. Later, leaf area was estimated as a percentage of the potential fully developed canopy of each tree. These visual estimates were supplemented by photography.

Canopy area normally increases as the leaves expand, but this may be checked by frost or by herbivores, so that the area may show a decline to a minimum. On the sampling occasion (usually the first or second week in June) when most of the trees have reached their minimum, their canopy leaf areas are recorded as leaf-area deficits. For example, a 25% deficit corresponds to a minimal area equivalent to 75% of full canopy. In many cases, this deficit represents incomplete expansion, as well as defoliation by caterpillars, but in the case of tree Q4, which has rarely produced more than 20% of its potential by this time, the values are for losses caused by defoliation alone.

From 1987 onwards, four water-filled plastic plant-pot saucers (diameter 20cm, depth 5cm) were pegged to the ground, one metre out from the trunk, around each of six of the trees (tree QE4 was excluded). At weekly intervals, caterpillars were removed and preserved from these frass traps, while faeces and small leaf fragments were filtered off, oven-dried and weighed, to provide an estimate of activity in the canopy.

Results

Leaf area deficits and flushing sequences

The maximum deficit for each tree in each year is plotted in Figure 1, which also shows the position of each tree in that year's flushing sequence . This sequence hardly varies from year to year: early flushers, typically Q6 and QE1, are usually followed within a few days by four others (eg Q2, Q8, Q5 and QE4), while Q4 is consistently last, its buds bursting three weeks or more after the early trees. Figure 1 also includes the mean deficit for each year, based on five trees (excluding Q4 and QE4). The most striking feature of these results is that mean deficits between 1982 and 1990 ranged from 30-78%, whereas since 1991 these values have generally been well below 10%. This contrast is also illustrated by the photographs of Q2 in May 1990 and 1994 (Figure 2).

During the first period, in the five years when mean deficits were less than $c50\%$, deficits were generally greatest on late or intermediate flushing trees. This is shown most clearly in 1982, when the deficits varied inversely with the values recorded for canopy expansion on 14 May, which ranged from 20% for QE4 and Q8 to over 70% for QE1 and Q2. During 1987 and 1989 highest deficits were recorded on intermediate flushers, while 1988 and 1989 provided the first exceptions to the view that early flushers escape defoliation: in both years, Q6 was badly affected, but not QE1, although both trees flushed at about the same time. Turning to the two years of greatest devastation, in 1984 the six trees were reduced to a similar extent, and in 1990 the two earliest flushers (QE1 and Q2) were among the most depleted.

In 1991, Q6 was the first to flush, and suffered the highest deficit, but thereafter individual values are too low to detect any obvious trends. Q4 escapes virtually unscathed every year, although there was appreciable damage in 1989 and 1990, the latter involving removal of most of the earliest foliage. Typically this tree produces isolated clumps of particularly large leaves, and achieves its maximal leaf area as much as two months after the first of the other trees, eg the end of July in 1987.

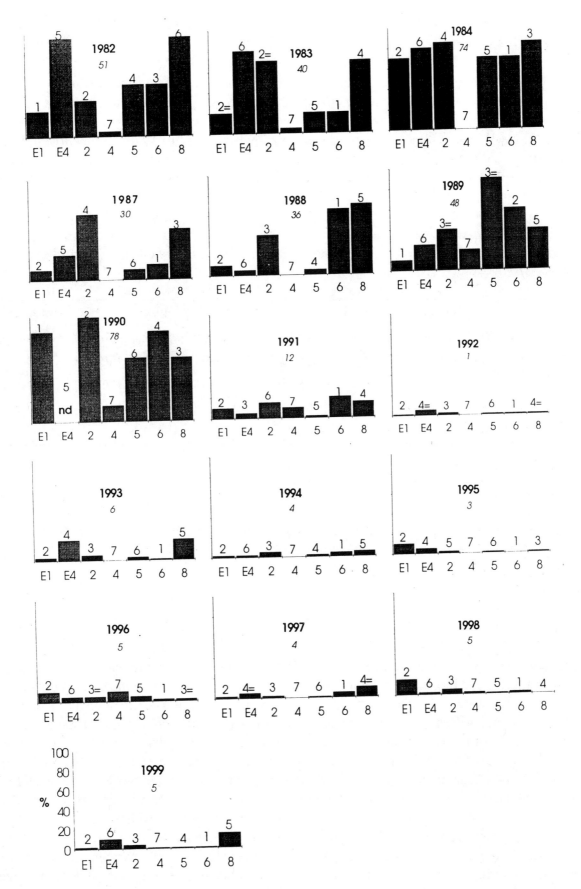

Figure 1. Maximum defoliation (leaf-area deficit) as a % of full canopy area for seven oaks at Chaddesley, 1982-99. Figures in italic are annual mean deficits based on five trees. Figures above each column indicate flushing sequence (1 = earliest)

Caterpillar numbers

Caterpillars from the frass traps under six of the trees were separated into geometrids, other Lepidoptera, and sawflies. The numbers in Figure 2 refer to the three consecutive weeks with the highest totals, a period which usually ended a week or so before the time of the greatest leaf-area deficit(Table 1). These three-week totals rose steadily from 1987 to 1990, largely driven by the geometrids, although other Lepidoptera made up a higher proportion in 1990 than in previous years. Since then, totals have fluctuated between $c200$ and 300, and in most years geometrids have been outnumbered by other Lepidoptera.

Table 1. Three-week periods of peak caterpillar numbers in frass traps under 6 oaks at Chaddesley and, in parentheses, dates when maximum leaf-area deficits were recorded, 1982-1999.

1982	---	(3 June)	1983	---	(13 June)	1984	---	(4 June)
1987	13/5-3/6	(10 June)	1988	11/5-1/6	(15 June)	1989	8/5-1/6	(8 June)
1990	3/5-24/5	(24 May)	1991	22/5-12/6	(12 June)	1992	15/5-11/6	(19 June)
1993	13/5-7/6	(7 June)	1994	24/5-14/6	(14 June)	1995	17/5-7/6	(26 June)
1996	30/5-20/6	(26 June)	1997	13/5-3/6	(3 June)	1998	12/5-3/6	(10 June)
1999	11/5-2/6	(9 June)						

Frass weights

Because of changes in procedure (see Discussion) detailed results are not presented here, but those for 1987-91 are summarized in Harding (1992).

Discussion

Methodology

The monitoring methods are somewhat crude. Partially, this was to ensure that monitoring could be easily performed by others, including project students, but also so that the various operations could be carried out within the constraints of a teaching timetable. The time taken to empty the frass traps could in itself have been used as an estimate of caterpillar activity during the preceding week. Given more time, the number of trees studied could usefully have been at least doubled. The flushing sequence could have been made more quantitative, eg in relation to the proportion of buds flushed (Hunter 1992).The estimates of canopy area were made by eye, and are liable to considerable error, so that the values in Figure 1 may be out by 10% or more. However, they indicate the relative condition of the various trees within a year, as well as major differences between years and between decades. The technique is also similar to that used by the Forestry Commission in their annual monitoring of forest condition (Innes 1990), involving estimating the percentage reduction in crown density compared to an 'ideal' tree or to a local reference tree. Photographs could have shown more close-up detail, and ideally each shot should have been taken from a fixed point; this proved impossible, however, because of the growth of birches and planted oaks.

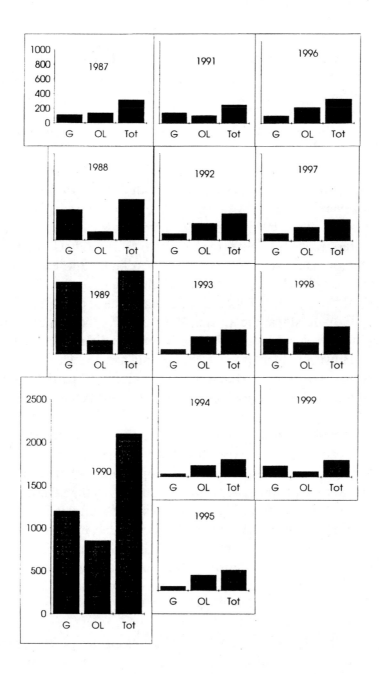

Figure 2. Numbers of caterpillars in frass traps during the three weeks of peak activity (April/May 1987-99). G = Geometrids, OL = other Lepidoptera, Tot = total including sawflies

The positions of the frass traps could have been randomized between sampling occasions. With the exception of leaf-rollers, such as *Tortrix*, they gave a reasonable estimate of populations of caterpillars in the canopy, although more effort should have gone into identifying them to species. Frass was separated by different means in different years, more recently by decanting and sieving, so that the dry weights include varying amounts of leaf fragments, as well as faeces; however, they are still of use when comparing activity between trees in a particular year.

Flushing sequence and its relation to herbivore activity

The fact that the flushing sequence of these seven oaks at Chaddesley shows so little variation from year to year is in agreement with observations on oaks at Roudsea Wood, Cumbria (Satchell 1962), at Wytham Wood (Gradwell 1974), in Silwood Park (Crawley & Akhteruzzaman 1988) and in woodlands in Belgium (Van Dongen *et al.* 1997). Such constancy is, presumably, largely a reflection of genetic variation between the individuals. At Silwood the date of budburst differed by about four weeks between the earliest and latest of 36 oaks, comparable to the situation at Chaddesley.

Elton (1966), on the basis of Satchell's (1962) work with *Tortrix viridana*, and of studies at Wytham, stated that severe defoliation was most likely to occur when emergence of the caterpillars from the egg virtually coincided with the flushing of a particular tree. Earlier flushing gave the leaves a good start, so enabling them to grow away from an attack, whereas young larvae were unlikely to survive on late flushing trees with closed buds. Subsequently, from further studies at Wytham, especially by Hunter (1992) and Hunter and West (1990), it was concluded that early flushers carry the most caterpillars and suffer the greatest damage. However, Crawley and Akhteruzzaman (1988) found that the average defoliation of their 36 trees over seven years was not correlated with the flushing order, while early flushers did not have significantly higher levels of defoliation. However, their latest tree was always the least damaged (cf. Q4 at Chaddesley). The differences between Wytham and Silwood may be partially related to the fact that only half of the 36 trees at Silwood are in woodland, the rest being free-grown in parkland.

At Chaddesley, this relationship showed considerable variation between years during 1982-90 (Figure 1). In 1982, 1983 and 1987 the two earliest flushers were among the least affected, but were among the worst in 1984 and 1990. The contrasting fortunes of QE1 and Q6 in 1988 and 1989 indicate that there is more to apparency (*sensu* Feeny 1976) than the flushing order, since both trees flushed at about the same time, but only Q6 was badly affected. This aspect might be affected by the species composition of the herbivores on each tree, eg the proportion that was of a highly mobile species, such as *T. viridana*, compared to the more parochial winter moth *Operophtera brumata*. In the latter the female is flightless, usually ovipositing on the tree on which she developed, which may increase the chances of synchrony between budburst and hatching of her eggs (Van Dongen *et al.* 1997).

During 1990 it seems that flushing at Chaddesley may have been checked by frosts in the second week of April. Damage was first noticed soon after early trees such as QE1 and Q2 flushed at the end of April, and the frass traps filled up to record levels during the first three weeks of May. This activity peak was the earliest recorded (Table 1), being possibly a couple of weeks ahead of the mass defoliation of 1980. Data for Q2 in 1990 illustrate what happens when there is insufficient foliage to support a tree's herbivore burden: relatively little frass, but large numbers of small, hungry caterpillars in the traps. The slightly earlier QE1 maintained enough foliage to grow away.

State of the canopy

Between 1982 and 1990, the annual mean canopy deficit at Chaddesley ranged from 30-78% (Figure 1), comparable to the average of 40% defoliation quoted for oaks at Wytham by Hunter and West (1990). The mean reduction in crown density for oaks from 85 sites throughout Britain, monitored during July and August for the Forestry Commission's Tree Health Survey, was *c*35% during 1987-98 (Redfern, Boswell & Proudfoot 1999).

92

Since 1991, deficits have been appreciably lower, generally well below 10%, corresponding to the situation in Alice Holt Forest, where healthy green canopies have been the norm following complete defoliation by *Tortrix* in 1981 (Winter, *pers. comm.*, 1995), and in Białowieża Forest, eastern Poland, where no damage has been reported in the foliage for the last four years (Jaroszewicz *pers. comm.* 1999). On the other hand, the Forestry Commission's Tree Health Survey results for 1998 include crown density reductions for oak of *c*35% in the western Midlands, and of 50-60% in Northumbria and parts of central Scotland (Redfern, Boswell & Proudfoot 1999). Derek Redfern (*pers. comm.)* reported that several oak sites in Scotland were completely defoliated in 1999, while Mick Crawley (*pers. comm.*) nominated Spinningdale as a site where oaks were stripped in five years out of nine. Scotland is also well known for the depredations of winter moth on Sitka spruce and heather (Kerslake *et al.* 1996).

The healthy state of the canopy at Chaddesley during this decade correlates with the dearth of caterpillars following their 1990 peak. A number of other long-term monitoring sites provide evidence of peaks and troughs in caterpillar abundance (Figure 3). In Hungary, where a Forest Damage Monitoring System has been in operation since 1962, the area of forest damaged annually by various caterpillars, including geometrids, has been correlated with drought indices, and the increased frequency and severity of drought in the last two decades is thought to have played a key role in provoking serious outbreaks of various major lepidopteran pests (Csóka 1994, 1997).

Outbreak years and climate change

Winter moth *Operophtera brumata* belongs to a group of a dozen or so species of forest Lepidoptera in N. America and Europe whose populations, at least in certain parts of their range, exhibit remarkably regular fluctuations, with major peaks at intervals of about six to eleven years. An example of such cyclic dynamics is provided by records for 1862-1968 of defoliation of mountain birch in Fennoscandia by the autumnal moth *Oporinia=Epirrita autumnata* and species of *Operophtera* including winter moth, where major outbreaks cycled with a periodicity of nine years (Haukioja *et al.* 1988). These dynamics are markedly influenced by negative feedback involving time-lags, commonly referred to as delayed density dependence or second-order feedback processes. Such processes could include combinations of host-plant quality varying with the degree of defoliation in the previous year(s), variation in lepidopteran quality (mediated either through physiology or genetics) or the influence of predators, parasites or diseases (Myers 1988; Berryman 1996; Roland 1990, 1994; Roland & Embree 1995).

At Wytham long-term work on the population dynamics of the winter moth has shown that the most likely cause of cycling is time-lagged pupal predation (Hunter, Varley & Gradwell 1997) whereas year-to-year fluctuations are particularly influenced by levels of winter disappearance between the egg stage and the descending pre-pupae. There appears to be strong selection pressure for synchronisation of egg hatch and bud-burst of the oaks, so that the larvae can make most use of the leaves at their most nutritious stage (Feeny 1970; Van Dongen *et al.* 1997; Varley, Gradwell & Hassell 1973).

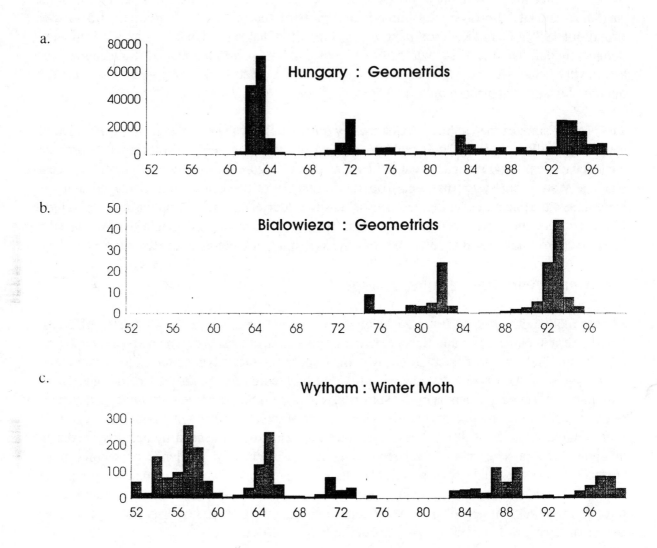

Figure 3. Long term monitoring of geometrid caterpillars. (A) Area of Hungarian forests defoliated (Csóka pers. comm.); (b) index of abundance of geometrids on hornbeam in late May, Białowieza Forest, Poland (Jaroszewicz pers. comm.); (c) abundance of *Operophtera brumata* (mean no m^{-2}) at Wytham Woods, (Cole pers. comm.)

Climate change might disturb the relationship between the timing of budburst in oaks and egg hatch of species such as winter moth. Studies by Buse and Good (1996), using elevated temperatures (+3°C) in experimental Solardomes, showed that this synchronization was generally maintained. However, these could still be consequences for the breeding success of certain insectivorous birds, such as the great tit *Parus major* (McCleery & Perrins 1998). Ideally, egg laying in this species is timed so that when the young birds are about one week old, and requiring maximum feeding, the biomass of canopy caterpillars will be at its peak (Perrins 1991). This peak is normally taken to coincide with the half-fall date, ie when half of the winter moths have descended from the canopy to pupate. Largely because of the differential effects of higher temperatures on the rate of development of the caterpillars and on the incubation period of the hen bird, this synchronization is more likely to be upset by warmer than by colder summers (Van Noordwijk, McCleery & Perrins 1995).

In the Netherlands, Visser *et al.* (1998) found no evidence of an increase in the mean temperature for 1 March - 15 April since 1973, nor in the lay date at the Hoge Veluwe, in contrast to a clear trend since 1970 towards earlier lay dates at Wytham (McCleery & Perrins 1998). However, mean temperatures for the subsequent 30 days had increased , inducing an advance of about nine days in the predicted date for peak caterpillar biomass, and intensifying selection for early laying.

Conclusions

Only by long-term studies can these interrelationships of plant - insect - bird and climate be elucidated, nor is there any guarantee as to when the next big defoliation year will occur. At Chaddesley it should be May 2000. Maintaining routine observations in the face of this uncertainty is one of the challenges for woodland ecologists.

Acknowledgements

I should like to thank the late John Cadbury, warden John Robinson (NCC/English Nature) and Harry Green and Helen Woodman (Worcs. Wildlife Trust) for permission to work at Chaddesley. I am most grateful to the following for supplying me with data, and for discussions: Alan Buse, Lionel Cole, Mick Crawley, Gyögy Csóka, Bogdan Jaroszewicz, Chris Perrins, Derek Redfern, Jens Roland, Pal Szontagh, Rosemary Winnall and Tim Winter. Robin Stuttard patiently computerized my diagrams. Finally, my thanks to the British Ecological Society for a Small Ecological Project grant to support this monitoring into the next millennium.

References

BERRYMAN, A.A. 1996. What causes population cycles of forest Lepidoptera? *Trends in Ecology & Evolution*, **11**, 28-32.

BUSE, A. & GOOD, J.E.G. 1996. Synchronization of larval emergence in winter moth (*Operophtera brumata* L.) and budburst in pedunculate oak (*Quercus robur* L.) under simulated climate change. *Ecological Entomology*, **21**, 335-343.

CRAWLEY & AKHTERUZZAMAN, M. 1988. Individual variation in the phenology of oak trees and its consequences for herbivorous insects. *Functional Ecology*, 2, 409-415.

CSÓKA, G. 1994. (In Hungarian) [Damage of Lepidoptera in Hungarian forests between 1961 and 1993]. *Novenyvedelem*, **30**, 263-268.

CSÓKA, G. 1997. Increased insect damage in Hungarian forests under drought impact. *Biologia, Bratislava*, **52**, 159-162.

ELTON, C.S. 1966. *The pattern of animal communities*. London, Methuen.

FEENY, P. 1970. Seasonal changes in oak leaf tannins and nutrients as a cause of spring feeding by winter moth caterpillars. *Ecology*, **51**, 565-581.

FEENY, P. 1976. Plant apparency and chemical defence. *Recent Advances in Phytochemistry*, **10**, 1-40.

GRADWELL, G. 1974. The effects of defoliators on tree growth. *In:* M.G. MORRIS & F.H. PERRING, eds. *The British Oak*, pp 182-193. Faringdon, E.W. Classey.

HARDING, D.J.L. 1992. Defoliation patterns amongst Chaddesley oaks. *In:* J.R. PACKHAM & D.J.L. HARDING, eds. *Woodland establishment, maintenance and assessment*. pp 69-77. Wolverhampton, Woodland Research Group.

HAUKIOJA, E., NEUVONEN, S., HANHIMAKI, S. & NIEMELA, P. 1988. The autumnal moth in Fennoscandia. *In:* A.A. BERRYMAN, ed. *Dynamics of forest insect populations*. pp163-178. New York, Plenum Press.

HUNTER, M.D. 1992. A variable insect-plant interaction: the relationship between tree budburst phenology and population levels of insect herbivores among trees. *Ecological Entomology*, **16**, 91-95.

HUNTER, M.D., VARLEY, G.C. & GRADWELL, G.R. 1997. Estimating the relative roles of top-down and bottom-up forces on insect herbivore populations: a classic study revisited. *Proceedings of the National Academy of Sciences, USA*, **94**, 9176-9181.

HUNTER, M.D. & WEST, C. 1990. Variation in the effects of spring defoliation on the late season phytophagous insects of *Quercus robur*. *In:* A.D. WATT, S.R. LEATHER, M.D. HUNTER & N.A.C. KIDD, eds. *Population dynamics of forest insects*. pp 123-134. Andover, Intercept.

INNES, J.L. 1990. *Assessment of tree condition*. London, HMSO. Forestry Commission (Field Book 12).

KERSLAKE, J.E., KRUUK, L.E.B., HARTLEY, S.E. & WOODIN, S.J. 1996. Winter moth (*Operophtera brumata* (Lepidoptera: Geometridae)) outbreaks on Scottish heather moorlands: effects of host plant and parasitoids on larval survival and development. *Bulletin of Entomological Research*, **86**, 155-164.

McCLEERY, R.H. & PERRINS, C.M. 1998. ... temperature and egg-laying trends. *Nature*, **391**, 30-31.

MYERS, J.M. 1988. Can a general hypothesis explain population cycles of forest insects? *Advances in Ecological Research,* **18**, 179-242.

PACKHAM, J.R., HARDING, D.J.L., HILTON G.M. & STUTTARD, R.A. 1992. *Functional ecology of woodlands and forests.* London, Edward Arnold.

PERRINS, C.M. 1991. Tits and their caterpillar food supply. *Ibis,* **133**, (Suppl.1) 49-54.

REDFERN, D., BOSWELL, R. & PROUDFOOT, J. 1999. *Forest condition 1998.* Edinburgh, Forestry Commission (Information Note 19).

ROLAND, J. 1990. Interaction of parasitism and predation in the decline of winter moth in Canada. *In:* A.D. WATT, S.R. LEATHER, M.D. HUNTER and N.A.C. KIDD, eds. *Population dynamics of forest insects,* pp 289-302. Andover, Intercept.

ROLAND. J. 1994. After the decline: what maintains low winter moth density after successful biological control? *Journal of Animal Ecology,* **63**, 392-398.

ROLAND J. & EMBREE, D.G. 1995. Biological control of the winter moth. *Annual Review of Entomology,* **40**, 475-492.

SATCHELL, J.E. 1962. Resistance in oak *Quercus* spp to defoliation by *Tortrix viridana* L. in Roudsea Wood National Nature Reserve. *Annals of Applied Biology,* **50**, 431-442.

VAN DONGEN, S., BACKELJAU, T., MATTHYSEN. E. & DHONDT, A.A. 1997. Synchronization of hatching date with budburst of individual host trees (*Quercus robur*) in the winter moth (*Operophtera brumata*) and its fitness consequences. *Journal of Animal Ecology,* **66**, 113-121.

VAN NOORDWIJK, A.J., McCLEERY, R.H. & PERRINS, C.M. 1995. Selection for the timing of great tit breeding in relation to caterpillar growth and temperature. *Journal of Animal Ecology,* **64**, 451-458.

VARLEY, G.C., GRADWELL, G.R. & HASSELL, M.P. 1973. *Insect population ecology.* Oxford, Blackwell Scientific Publications.

VISSER, M.E., VAN NOORDWIJK, A.J., TINBERGEN, J.M. & LESSELLS, C.M. 1998. Warmer springs lead to mistimed reproduction in great tits (*Parus major*). *Proceedings of the Royal Society of London B,* **265**, 1867-1870.

THE LONG-TERM PHENOLOGY OF WOODLAND SPECIES IN BRITAIN

Tim Sparks

NERC Institute of Terrestrial Ecology, Monks Wood, Abbots Ripton, Huntingdon, Cambridgeshire PE17 2LS

Summary

With records going back over 250 years, phenology may provide the longest *written* biological record in Britain. Current phenological records may be collected by scientific institutes as part of a wider remit or be the pastime of individuals. Historic data exist in abundance but, despite a current initiative, some may never surface from obscurity. This paper examines data on woodland species from a range of past and current sources and examines how changes in timing of a number of different taxa relate to temperature.

Introduction

Phenology is described by The Concise Oxford Dictionary as the "Study of the times of recurring natural phenomena esp. in relation to climatic conditions". It has a long history (Clarke 1936). In Japan and China, some of the dates of flowering of cherry and peach trees associated with festivals exist from the eighth century. In Britain we know of records dating back to 1736 and, sooner or later, I would expect we shall uncover some older data. Many people are aware of the records kept by Gilbert White, an obscure Hampshire clergyman who just happened to have a publisher for a brother! At the end of the eighteenth century it was possible for a gentleman to purchase a printed naturalists' diary in which space was designated for recording weather conditions and special biological observations. From 1875 a network of recorders was co-ordinated by the Royal Meteorological Society (RMS) who wished to examine the relationship between meteorological events and the natural world. This network failed to continue after 1947 when the RMS could not find a body willing to take on co-ordination of the scheme. Many other schemes have existed at some point in time; for example, the Natural History Journal, and the Malborough Natural History Society. Some of these were incorporated within the larger RMS scheme. For most of the current century the British Naturalists Association has managed to maintain a phenological scheme. In recent times the long term records of individuals have come to light (eg Fitter *et al.* 1995; Sparks, Carey & Combes 1997). Phenological data is also available from various monitoring schemes such as the Wytham Wood study, the Rothamsted Insect Survey, the British Trust for Ornithology schemes, and the Butterfly Monitoring Scheme. In this paper I use examples of the timing of woodland species from a range of sources to show how they have related to temperature in the past. English names are used for the common species but there is a full listing of their scientific names in the Appendix.

Data sources and results

Marsham data. Robert Marsham FRS started his collection of "Indications of Spring" in 1736 and continued until his death in 1798. Marsham was an early scientist, dedicated to improving the profitability of his estate in Norfolk by improving timber production. He undertook *replicated* experiments at a time when such things were rare and his longevity ensured that he was able to see the outcome of many of his experiments on, for example, root cutting, trenching and bark scrubbing. In my opinion he deserves our attention every bit as much as Gilbert White. Some of his "Indications of Spring" are summarized in Table 1. The unique aspect of Marsham's record is that it was continued by successive generations of his family. I use the Central England Temperature (CET) for comparison throughout this paper. For convenience this table just examines the relationship with first quarter temperature. We reported results until 1947 (Sparks & Carey 1995), but since then I have acquired further data down to 1958.

Table 1. A summary of the timing of tree leafing dates recorded by Robert Marsham FRS and his successors. Means are presented for the parts of the three centuries covered by the record. Dates have been regressed on mean January-March CET; a negative response (s.e. in brackets) indicating the number of days earlier for a 1°C rise in temperature.

Event Leafing of:	number of years of data	18th century mean (1736-1799)	19th century mean (1800-1899)	20th century mean (1900-1958)	Response to JFM mean temperature
Hawthorn	184	March 14	March 6	March 12	-9.9 (0.8)
Sycamore	175	April 4	April 4	March 27	-6.7 (0.7)
Birch	175	April 2	April 4	April 11	-5.2 (0.7)
Elm	148	March 31	April 8	April 4	-5.7 (0.9)
Mountain ash	165	April 5	April 8	April 7	-5.6 (0.6)
Oak	178	April 28	April 24	April 21	-5.6 (0.5)
Beech	178	April 25	April 20	April 19	-3.0 (0.4)
Horse chestnut	173	April 4	April 4	April 5	-4.8 (0.5)
Sweet chestnut	155	April 23	April 18	April 18	-5.5 (0.6)
Hornbeam	160	April 7	April 5	April 13	-6.1 (0.8)
Ash	158	April 29	April 30	April 26	-3.5 (0.7)
Lime	167	April 15	April 15	April 11	-5.2 (0.6)
Field maple	115	April 17	April 15	April 22	-4.4 (0.9)
mean JFM temperature		3.8	4.1	4.5	

There is no evidence to support the "oak before ash" rhyme in these data; a correlation of 0.03 exists between early summer rainfall and the difference between oak and ash leafing. Interestingly, there are also rhymes concerning these two species from both Norway and Germany. If anything, oak has been getting progressively earlier whilst ash timing has been more static; as one might expect by examining their respective responses to temperature. The leafing species appear to fall into three broad groups: hawthorn with a strong response of c 10 days, the majority of species with a response of 5-6 days, and those with a lower response (ash, beech and possibly field maple).

Royal Meteorological Society data. A summary of the flowering records from the RMS phenological scheme were given by Jeffree (1960) who reported national mean flowering dates over a 58 year period. The response of four selected species has been summarized in Table 2. These national mean data show a similar level of response to the single site Marsham leafing data. One obvious feature of the data is the lateness of the spring flowering species (hazel, wood anemone) in the 1940's which appears to be associated with a series of cooler springs (1940 2.4°C, 1942 2.1°C, 1947 1.3°C). Of some surprise to me is the way that ivy flowering responds to temperature. My empirical research has generally shown that it is the spring species that are most responsive to temperatures and the summer ones less so. Here however, ivy shows a highly significant response to temperature. To show more clearly how these national data relate to temperature the responses of hazel and ivy are shown in Figure 1.

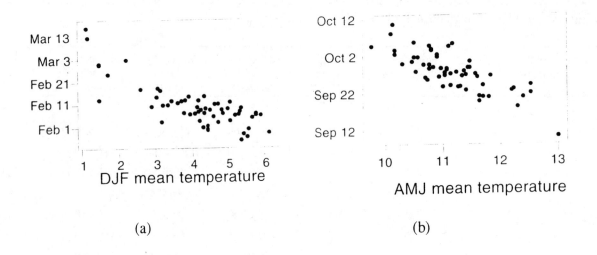

(a)　　　　　　　　　　　　　　　　　　(b)

Figure 1. The response of (a) hazel to Dec/Jan/Feb mean temperatures) and (b) ivy flowering times to April/May/June temperatures. Both relationships are highly significant (p<0.001) with r^2 of 65% and 63% respectively.

Table 2. A summary of the response of the flowering of certain species to temperatures of three month periods. Data are British Isles annual means.

Species	1890s	1900s	1910s	1920s	1930s	1940s	comparison with months	response (se)
Hazel	Feb 9	Feb 11	Feb 9	Feb 7	Feb 8	Feb 18	DJF	-6.1 (0.6)
Wood anemone	Mar 28	Mar 31	Mar 29	Mar 25	Mar 30	Apr 1	JFM	-4.8 (0.4)
Hawthorn	May 12	May 19	May 13	May 11	May 14	May 10	MAM	-9.6 (0.8)
Ivy	Sep 27	Oct 2	Sep 24	Sep 30	Sep 29	Sep 24	AMJ	-6.4 (0.6)
JFM mean	4.2	4.3	4.5	5.2	4.8	3.9		
AMJ mean	11.2	10.7	11.3	10.8	11.3	11.7		

Another source of data from the RMS was reported by Sparks and Yates (1997). These data were potentially very useful in showing the relationship between first emergence of orange tip butterflies and the timing of one of its foodplants, garlic mustard. The data suggests that the two species remain in synchrony over a wide range of temperatures (Figure 2). This is a very important feature, suggesting that synchrony may be maintained under elevated temperatures. A similar synchrony was shown for winter moth and oak trees by Buse and Good (1996) under experimental conditions.

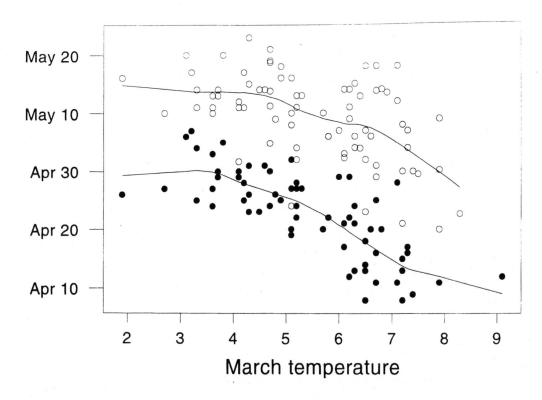

Figure 2. The relationship between garlic mustard first flowering times (solid circles) and orange tip butterfly first appearance times (open circles). Data are annual means for England and Wales taken from RMS series. A smoothed (LOWESS) curve has been superimposed.

Butterfly Monitoring Scheme. The timing of certain butterfly species from the Butterfly Monitoring Scheme (BMS - Pollard & Yates 1993) is reported here. Recent work (Roy & Sparks in press) has shown that the timing of most butterfly species responds to increased temperatures. The response varies between species but appears to be of the same order of magnitude as the tree data reported above. Figure 3 shows the timing of mean UK first appearance and mean peak appearance of the ringlet for the years 1976-1998. Appearance in 1976 was particularly early. What is apparent from this graph is that the ringlet has got progressively earlier in recent years and that peak timing mirrors first appearance. The butterfly responds well to early summer temperatures and appears capable of advancing its appearance by about seven days per degree Celsius warming.

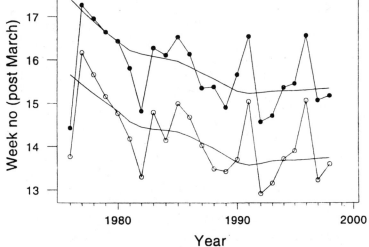

Figure 3. UK mean first appearance (open circles) and peak timing (solid circles) of the ringlet butterfly. The vertical axis represents weeks (BMS begins in the first week of April). A smoothed (LOWESS) line has been superimposed.

Combes Phenological Record Jean Combes has kept a summary of the leafing of oak, ash, lime and horse chestnut (Sparks, Carey & Combes 1997). This record, starting in 1947, is now invaluable to examine the post-war period when changes to temperature are thought to be most pronounced. Table 3 shows that the decadal mean first leafing dates for the four species were all much earlier in the 1990s - the warmest decade on record. Figure 4 shows that the responses from both Marsham and Combes records to temperature are similar, ie that current response is very similar to the historical response. Figure 5 shows that there is a continued trend towards earliness in oak leafing data when the two records are amalgamated.

Table 3. The average leafing dates of four tree species in Surrey over the last five decades and average January-March mean temperature (°C).

Decade	1950-1959	1960-1969	1970-1979	1980-1989	1990-1999
Oak	Apr 30	Apr 26	Apr 24	Apr 20	Apr 10
Ash	May 6	May 7	May 7	Apr 30	Apr 24
Horse chestnut	Mar 25	Mar 27	Mar 31	Mar 27	Mar 15
Lime	Apr 7	Apr 9	Apr 12	Apr 14	Apr 3
JFM Temperature	4.3	4.2	4.5	4.3	5.6

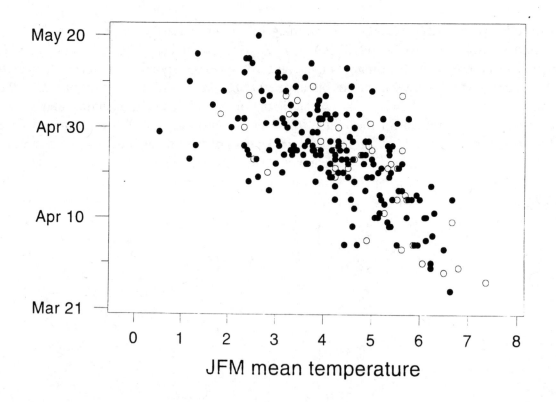

Figure 4. The relationship between oak first leafing date and January-March mean temperature for the Marsham (solid circles 178 years) and Combes (open circles 43 years) records.

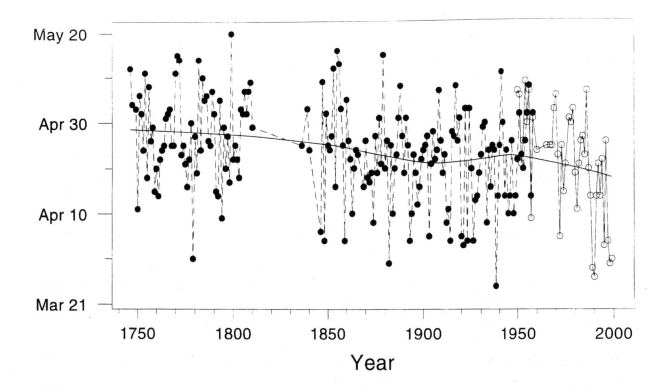

Figure 5. The oak leafing time series provided by the Marsham and Combes records. A smoothed (LOWESS) line has been superimposed.

Bird migratory timing There is growing evidence suggesting a response in the timing of migratory birds (Sparks 1999). Data on arrival times of migrants from 1960-1996 has been obtained from eight coastal observatories: Holme, Sandwich, Dungeness, Portland, Skokholm, Bardsey, Calf of Man and Walney. Two woodland species, chiffchaff and blackcap, appear to be getting earlier (Figure 6), and the response to temperature is in the order of 2-3 days per degree Celsius. These species are also increasingly overwintering in the UK, and even as far north as Norway.

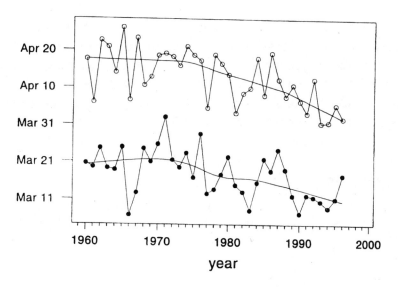

Figure 6. Annual arrival dates of the chiffchaff (solid circles) and blackcap (open circles) for the period 1960-1996. Data are averaged across eight coastal observatories.

Discussion

The results shown here indicate the level of response between temperature and the timing of various biological events associated with woodland plant species. Examples have been drawn from a range of taxa, both historic and current data and both local and national data sources. In general, phenology consists of times of first events, but other events also exist. Where these additional data do exist, for example butterflies, there appears to be a close link between first appearance and peak appearance, at least at a national scale. Other data, as yet unpublished, show a similar link between first leafing and full leafing dates of a large number of woody plant species. So, despite criticism, it appears that collecting first dates can be useful.

There is evidence that some species will remain synchronized with others on which they are reliant. However not everything in the garden may be rosy. The Wytham study (Perrins 1991) has suggested a possible mismatch between caterpillar timing and blue tit breeding success - so synchrony between vertebrates and invertebrates or plants may be less well matched.

Woodland plant communities are dynamic but may take several decades to change. If one species was able to take competitive advantage over another by dint of its earlier leafing then it might gradually become dominant. Some experimental work at Monks Wood has shown hawthorn to out-compete beech in hedges, but this situation may be reversed if beech were allowed to acquire a height advantage. Might we see ash-oak woodland (with ash dominant) replaced by oak-ash woodland (oak dominant) if oak is responding more to climate warming? Researchers in the Netherlands are looking at woodland community structure in phenological modelling, particularly with regard to the dangers of late spring frost damage.

Migrant bird species do not seem to be taking full advantage of the earlier leafing of trees; perhaps they need to remain cautious about adapting too quickly? Some species (e.g. oak, chiffchaff) appear to be getting earlier than we might expect by their response to air temperature alone. Perhaps we should be looking at additional information, e.g. on soil temperatures or European temperatures? Perhaps there are other causes we need to consider?

All of the above work shows that there is a wealth of data on the timing of woodland species. However, analysis of such data is at a very earlier stage and there is a danger of attempting to run before we can walk. We are getting there slowly. The identification and curation of data sources must be a priority at the moment. And finally a word of caution. Changes brought about by climate warming are likely to be subsidiary to those brought about by direct human activity, e.g. farmland and woodland management. However, climatic induced changes may be sufficient to tip the balance in fragile ecosystems.

Acknowledgements

I remain grateful to all those, past and present, who have collected the data reported here.

References

BUSE, A. & GOOD, J.E. 1996. Synchronisation of larval emergence in winter moth (*Operophtera brumata* L.) and budburst in pedunculate oak (*Quercus robur* L.) under simulated climate change. *Ecological Entomology*, **21**, 335-343.

CLARKE, J.E. 1936. The history of British phenology. *Quarterly Journal of the Royal Meteorological Society,* **62,** 19-23.

FITTER, A.H., FITTER, R.S.R., HARRIS, I.T.B. & WILLIAMSON, M.H. 1995. Relationships between first flowering date and temperature in the flora of a locality in central England. *Functional Ecology,* **9,** 55-60.

JEFFREE , E.P. 1960. Some long-term means from The Phenological Reports (1891-1948) of the Royal Meteorological Society. *Quarterly Journal of the Royal Meteorological Society,* **86,** 95-103.

PERRINS, C.M. 1991. Tits and their caterpillar food supply. *Ibis,* **133**(supplement), 49-54.

POLLARD, E. & YATES T.J. 1993. *Monitoring butterflies for ecology and conservation.* London, Chapman and Hall.

ROY, D.B. & SPARKS, T.H. in press. Phenology of British butterflies and climate change. *Global Change Biology.*

SPARKS, T.H. 1999. Phenology and the changing pattern of bird migration in Britain. *International Journal of Biometeorology,* **42,** 134-138.

SPARKS, T.H. & CAREY, P.D. 1995. The responses of species to climate over two centuries: An analysis of the Marsham phenological record, 1736-1947. *Journal of Ecology,* **83,** 321-329.

SPARKS, T.H., CAREY, P.D., & COMBES, J. 1997. First leafing dates of trees in Surrey between 1947 and 1996. *The London Naturalist,* **76,** 15-20.

SPARKS, T.H. & YATES, T.J. 1997. The effect of spring temperature on the appearance dates of British butterflies 1883-1993. *Ecography,* **20,** 368-374.

Appendix

English names of species and their scientific equivalents

Ash	*Fraxinus excelsior*	Horse chestnut	*Aesculus hippocastanum*
Beech	*Fagus sylvatica*	Ivy	*Hedera helix*
Birch	*Betula pendula*	Lime	*Tilia* x *vulgaris.*
Blackcap	*Sylvia atricapilla*	Mountain ash	*Sorbus aucuparia*
Chiffchaff	*Phylloscopus collybita*	Oak	*Quercus robur*
Elm	*Ulmus* spp.	Orange tip	*Anthocharis cardamines*
Field maple	*Acer campestre*	Ringlet	*Aphantopus hyperantus*
Garlic mustard	*Alliaria petiolata*	Sweet chestnut	*Castanea sativa*
Hawthorn	*Crataegus monogyna*	Sycamore	*Acer pseudoplatanus*
Hazel	*Corylus avellana*	Winter moth	*Operophtera brumata*
Hornbeam	*Carpinus betulus*	Wood anemone	*Anemone nemorosa*

A MONITORING PROGRAMME TO ASSESS CHANGES IN THE BALANCE BETWEEN SPRING VERNALS AND WARMTH-DEMANDING SPECIES IN THE GROUND FLORA OF A SUSSEX WOOD

M.C. Pilkington
Centre for Continuing Education, University of Sussex, Falmer, Brighton BN1 9RG

Introduction

The Plantlife report *Death Knell for Bluebells* (1991) suggested that warmer winters may lead to increasing vigorous and early growth of warmth-demanding species at the expense of spring vernals such as bluebells with the result that our woods in 50 years time will be filled with grass not bluebells. Many scientists will be unhappy with this sort of extrapolated prediction, particularly when expressed in rather polemic language, but Grime (1993) suggests that changes in timing of growth and the relative abundance of species of high and low DNA amounts are likely to provide a sensitive indicator of the first effects of climate change on existing plant communities. Barkham (1994) spells out the need to risk making such predictions and to follow on with the corresponding adjustments to our behaviour. In order for this to be effective we need to take the general public with us and an ideal way of doing this is to involve adults in the monitoring of representative plant communities in their local area. In Mid Sussex, Continuing Education natural history classes have provided a focus for such long-term monitoring studies. Knight (1995) stresses that students should understand the science behind the monitoring rather than simply acting as data collectors. Here students have been involved in the decision-making process at all stages.

Methods

A relatively undisturbed area of ash-maple (W8b) woodland (Rodwell 1991) in Hoe Wood (National Grid Reference TQ 218134 - 218135) was chosen at Woods Mill, the Sussex Wildlife Trust Countryside Centre. A series of six 1 m x 1 m quadrats was set out in 1993 at 5 m intervals along a footpath to avoid trampling large areas of bluebells. Quadrats were placed 30 cm in from the edge of the footpath on the western side. In 1995 a second series of 15 1 m x 1 m quadrats was set out in a similar way at 10 m intervals further along the footpath. The position of quadrats was adjusted to avoid large trees and other discontinuities.

Domin values were recorded for all the ground flora species in each quadrat when the vegetation reached the same stage in spring emergence each year. For series one this was when the wood anemone *Anemone nemorosa* leaves had just emerged. However, this timing was found to be very critical as anemone leaves go on emerging and opening out as the plants are flowering and so the second series was set up and monitored at a later time of year when all the spring vernal foliage was fully expanded and timing was therefore less critical. Groups of 3-4 students from natural history classes (Centre for Continuing Education, University of Sussex) worked on each quadrat and Domin values were checked for consistency by independent monitoring of some quadrats by two groups in turn.

As well as monitoring the woodland quadrats, students recorded the first appearance of key species in their local areas of Mid Sussex on Search for Springtime sheets which they helped to design.

Results

There was no evidence from the Search for Springtime sheets of changes in the relative timing of growth of species such as wood anemone and bluebell with high DNA amounts and species such as cow parsley with low DNA amounts over the period (1993-1999) studied. The pattern of emergence of bluebell leaves closely mirrored that of hazel *Corylus avellana* catkin maturation which is weather dependent and provided the earliest indication of an early or late season. The emergence of wood anemone leaves and the flowering of both wood anemone and bluebell varied from year to year in a remarkably consistent pattern which was also in line with the pattern for the slightly later flowering cow parsley.

The quadrat monitoring demonstrated the stability of the woodland ground flora from one year to the next. An average Domin value was calculated for each year (1993-1999) over the six quadrats in series one and for each year (1995-1999) for the 15 quadrats in series two. There was little change in the abundance of bluebell over the seven and five years of monitoring respectively, or in the abundance of cow parsley which it was assumed would be the major warmth demanding competitor. Apparent changes in the abundance of wood anemone in the first series of quadrats may have been due to the difficulty of getting the time of the monitoring right since in some years leaves continued to emerge and expand as plants were flowering.

Other species, such as celandine *Ranunculus ficaria*, goosegrass *Galium aparine*, stitchwort *Stellaria holostea*, moschatel *Adoxa moschatellina*, and grass species, showed more variation, but there was no consistent trend over the years. Principal Components Analysis of the average Domin values for each year for these species also showed no trend and removing species successively showed no improvement. If changes were taking place in the relative abundance of high and low DNA species such as analysis would be expected to show a trend over the years.

Quadrats in the second series were also sorted into those with cow parsley and those with little or no cow parsley and average Domin values calculated for each group separately. Any changes reflecting competition between high and low DNA species might be expected to show up first in the quadrats with the larger amounts of cow parsley, but again there was very little change in the abundance of bluebell and such change as there was followed the same pattern as in quadrats with little or no cow parsley.

Discussion

In the part of Hoe Wood studied there is no dogs mercury *Mercurialis perennis* and bluebell is the dominant species with abundant wood anemone in many of the quadrats. This is to be expected on the gentle slope of the wood where water cannot easily percolate away and the soil is not sufficiently aerated for *Mercurialis* (Rodwell 1991). Wood anemone grows only where water logging restricts the growth of bluebell since it is intolerant of shading and has leaves which emerge later in the year (Pigott 1982). Over the period of this study no significant changes have occurred in the abundance of bluebell and wood anemone. Similarly warmth-demanding competitor species such as cow parsley showed no consistent increase in abundance over the years. This agrees with the Search for Springtime record sheets which showed no evidence of changes in the relative timing of growth of plants with high and low DNA amounts.

The study then provides a relatively stable base line against which future changes in the ground flora can be readily assessed. It is intended that monitoring should continue on an annual basis.

As well as changes in the balance of spring vernals and warmth demanding species, these quadrats should give early indications of the significant effect that the increasing severity of summer drought is predicted (Barkham 1994) to have on bluebells and wood anemones particularly in the dry South East. The importance of studies where the response of plants to changes in their natural environment is looked at within the context of the vegetation of which they are a part was stressed by Pigott in 1982 and applies particularly to studies on the effect of climate change on both plants and animals. However, the long-term nature of such studies makes them unattractive to researchers under pressure to produce publications for the next Research Assessment Exercise.

The relatively small number of plants found in the ground flora of our woodlands and the ease with which they can be identified, makes monitoring exercises such as this eminently suitable for adult education classes. Concern about changes to the countryside which they know and love, provides the stimulus and context for serious engagement with unfamiliar concepts and students learn about scientific investigation through participation in the process. They are thus better equipped to understand the environmental issues raised by global warming and to be convincing advocates for sustainable life styles.

References

BARKHAM, J. 1994. Climate change and British wildlife. *British Wildlife*, **5**, 169-180.

GRIME, J.P. 1993. Vegetation functional classification systems as approaches to predicting and quantifying global vegetation change. *In*: SOLOMON, A.M. & SHUGART, H.H., eds. *Vegetation Dynamics and Global Change*, p. 293-305. London, Chapman & Hall.

KNIGHT, D. 1995. Participative environmental research and the role of continuing education. London, *SCUTREA Conference papers*. pp. 108-112.

PLANTLIFE. 1991. *Death Knell for Bluebells*. London, Plantlife.

PIGOTT, C.D. 1982. The experimental study of vegetation. *New Phytologist*, **90**, 389-404.

RODWELL, J.S. 1991. *British Plant Communities, Volume I. Woodlands and Scrub*. Cambridge, Cambridge University Press.

Search for Springtime 1999

		Date	Location notes
Hazel catkins with pollen		Jan 26	
Dogs Mercury	leaves	Dec 16 1998	S Downs - north facing slope from Underhill Lane car park to the Downs of adjoining track from Ditchling Beacon to Jack & Jill Windmills
	Flowers	Jan 16	
Cuckoo pint	leaves	Dec 14 1998	Footpath running north to south from edge of Folders Lane estate (east edge) to Folders Lane
	flowers	April 12	
Celandine flowers		Feb 23	
Nettle shoots 1 ft high		Feb 23	Bedelands-Long Wood/Coopers Shaw
Wood-anemone	leaves	Jan 30	Blackbrook Woods. South-east corner from Ditchling Common/Spatcham Lane
	flowers	Mar 8	
Pussy-willow with pollen		Mar 8	St Mary's-Ditchling Common
Blackthorn flowers		Mar 1	Footpath north to Folders Lane
Primrose flowers		Jan 14	Blackbrook Woods
Woodland Hawthorn	leaves	Mar 8	Ditchling Common - Woodland area facing south
	flowers	April 14	
Common hawthorn	leaves	Mar 8	
	flowers	April 10	
Bluebell	leaves	Dec 1 1998	Blackbrook Woods
	flowers	April 1	Blackbrook Woods
Goose-grass shoots 1 ft		Feb 23	Bedelands - Long Wood
Cow parsley flowers		April 12	Footpath from estate to Folders Lane
Jack-in-the-hedge flowers		April 19	
Hornbeam leaves open		March 28	Bedelands. Big Wood
Beech leaves open		April 5	Bedelands. Old Arable
Oak leaves open		N/R	- N/R
Ash leaves open		April 17	Lower Plantation
Ox-eye daisy flowers		May 3	Ditchling Common Car Park

LONG-TERM MONITORING OF THE HEALTH OF BEECH (*FAGUS SYLVATICA* L.) TREES OF ALL AGES IN SURREY

G.H. Stribley
Apple Tree House, Highfield Road, West Byfleet, Surrey KT14 6QT

Introduction

Long-term monitoring of beech trees at amenity sites in Surrey set up in 1989 included young trees and saplings. Decline was reported up to 1993 (Stribley 1996) and this report updates the results to 1998.

Methods

Details of the sites and methods were given previously (Stribley 1996). Results of canopy health assessments are reported for the number of trees >10 cm diameter at breast height (dbh). Nower Wood (n=81) and Farnham Park (n=30) were monitored from the winter of 1989 and Sheepleas (n=26) and Virginia Water (n=26) from the summer of 1991. Bark stripping scores were based on the extent of stripping related to the girth at breast height (Table 1). Roloff twig (RT) was the winter assessment of the twig pattern: score 0 was healthy while score 3 corresponded to marked deterioration. Crown density was assessed in summer for percentage loss of crown density: 0 = 0 -10; 1 = 11-25; 2 = 26-60; 3 = >60% loss. Die back was also assessed in summer as the percentage of upper canopy affected: 0 = none, 1 = 1-10; 2 = 11-30; 3 = >30% dieback.

Results

Bark removal by grey squirrels was seen on all sized trees. At Nower Wood Table 1 shows that the greatest damage occurred in trees of 3.5-20 cm dbh with 54 per cent dying or decrowned and a further 20 per cent had severe scores of 3 or 4.

All sites showed an increase in the percentage of trees with a higher RT score (Figure 1). Young trees also showed marked deterioration at all sites whereas such changes would not normally be expected until around 140 years old. There were increased crown density and die back scores at all sites. RT winter scoring detected underlying changes in health not always apparent using crown density summer scoring (Figure 1; Stribley 1996). RT canopy scores have been correlated with changes in growth patterns shown by twig analysis and therefore RT scores are a valuable indicator of health in standing trees. Twig analysis has demonstrated significant correlation of growth changes with soil moisture deficit or ozone pollution (Stribley & Ashmore 1999). These factors are likely to have been important at the Surrey sites. Premature decline in canopy health of young trees is of particular concern which, together with serious squirrel damage, could have a major impact on regeneration of beech woodland within a changing climate and pollution environment.

References

STRIBLEY, G.H. 1996. Monitoring of the health of beech (*Fagus sylvatica* L.) trees of all ages. *Quarterly Journal of Forestry*, **90**,114-120.

STRIBLEY, G.H. & ASHMORE, M.R. 2000. Quantitative changes in twig growth pattern of young woodland beech (*Fagus sylvatica* L.) in relation to climate and air pollution over ten years. *In*: KIRBY, K.J. & MORECROFT, M.D., eds. *Long-term studies in British Woodland,* pp 112-119. (English Nature Science No. 34.) Peterborough, English Nature.

Table 1. Grey squirrel bark stripping at Nower Wood 1989-98: 2, 3, 4 represent increasing levels of squirrel damage.

Size class	dbh	Total no.	Bark stripped score				Number of trees (percentage of total)		
			2	3	4	2+3+4	Dead	Decrowned	Evidence of rotting
0	<3.5	45	2	1	0	3 (6.7)	4 (8.9)	0	3 (6.7)
1	3.5-10	21	3	6	1	10 (47.6)	9 (42.9)	1 (4.8)	5 (23.8)
2	11-20	20	0	4	5	9 (45.0)	5 (25.0)	7 (35.0)	9 (45.0)
3	21-30	11	2	6	0	8 (72.7)	0	2 (18.2)	8 (72.7)
4	31-40	6	2	1	0	3 (50.0)	0	0	3 (50.0)
1-2	3.5-20	41	3	10	6	19 (46.3)	14 (34.1)	8 (19.5)	14 (34.1)
0-4	<3.5-40	103	9	18	6	33 (32.0)	18 (17.5)	10 (9.7)	28 (27.2)

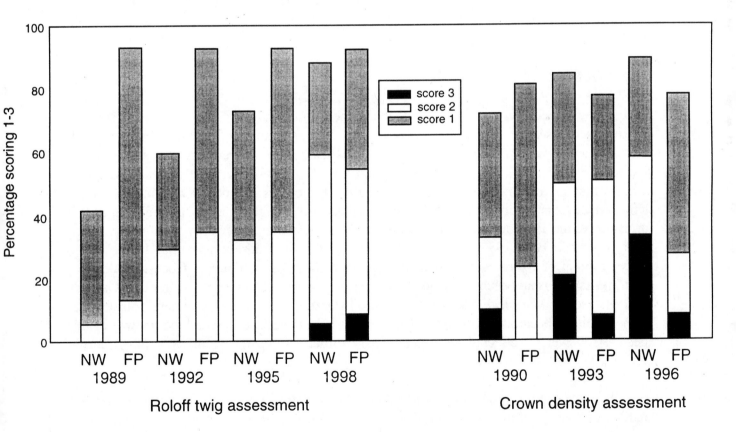

Figure 1. Canopy health of Nower Wood (NW) and Farnham Park (FP) trees 1989-98. Percentage of trees with score indicated by type of shading, for winter Roloff twig or summer crown density assessments.

111

QUANTITATIVE CHANGES IN TWIG GROWTH PATTERN OF YOUNG WOODLAND BEECH (*FAGUS SYLVATICA* L.) IN RELATION TO CLIMATE AND AIR POLLUTION OVER 10 YEARS

G.H. Stribley[1] and M.R. Ashmore[2]

[1] Apple Tree House, Highfield Road, West Byfleet, Surrey KT14 6QT
[2] Department of Environmental Science, University of Bradford, West Yorkshire BD7 1DP

Summary

Quantitative analysis of twig growth patterns (TA) from the upper canopy of 40 year old trees in two compartments at Wytham Wood near Oxford, was carried out for the growth years 1987-1996. Annual growth of leader or primary shoots, and lateral or secondary shoots, and the total number of subsidiary shoots were suppressed in drought years, while the percentage of acute shoots (growing at 40° or less to the parent stem) increased. In one compartment, these responses were significantly correlated with soil moisture deficit, while in the other both soil moisture deficit and ozone exposure were correlated with growth changes. These results support the interpretation that drought and ozone pollution are important factors in the long term decline of beech trees in southern Britain that has been observed using the Roloff winter twig pattern assessment (RT). There was good correlation between the RT canopy score and the percentage of acute shoots, supporting the use of the RT winter assessment to monitor beech tree health in the countryside. The TA methods can be valuable in examining detailed tree responses to a changing climate and pollution environment.

Introduction

Since 1989, a decline in canopy health of Surrey beech trees has been recorded using summer crown density assessment (score 0-3), which is affected by year to year factors, and winter Roloff twig (RT) assessment of twig pattern (score 0-3), which reflects longer-term changes on crown structure (Stribley 2000). Power, Ashmore and Ling (1995) also demonstrated an underlying decline of mature beech trees using the latter method at 16 sites in southern Britain. The causes of these declines in vitality are uncertain, but it is likely that drought years are a significant factor; for example, the 1976 drought caused long term suppression of leader growth in mature beech trees (Lonsdale *et al*.1989; Stribley 1993; Power 1994). Another potential factor is the concentrations of ozone, a pollutant associated with warm summers. Open-top chamber experiments exposing beech seedlings and saplings to ozone (Davidson, Ashmore & Garretty 1992; Pearson & Mansfield 1994) or unfiltered air (Durrant *et al*. 1992) have demonstrated that ozone, at the concentrations found in southern Britain, affects physiology and growth of this species. However, the effect of ozone in the field is difficult to separate from that of other climatic factors.

As an alternative to the visual assessment methods, such as RT, used in field monitoring, detailed analysis of twig growth patterns (TA) of twigs taken from the upper canopy can assist in understanding the relationship between growth patterns and environmental stresses. TA of Surrey trees has demonstrated a close correlation with the RT canopy score (Stribley 1996). At Wytham Woods, Oxfordshire, young trees had early stages of decline with RT scores of 1 and some on the borderline of ½, similar to the trees at five Surrey sites. The present study in Wytham Woods investigated whether TA could separate the effects of drought and ozone pollution on young woodland beech trees in two compartments with contrasting soil types.

Methods

Tree selection - Nine plantation trees of around 40 years old were selected which had RT scores of 1 or on the 0/1 borderline, in each of two woodland compartments. In compartment 55C, the soil was calcareous clay of less than 30 cm depth and the trees were in widely spaced rows, while in compartment 25C the deep soil was a variable sandy/clay loam and the trees were within a few metres of each other. The trees were felled as part of thinning management and two branches of about 5m were collected per tree for twig analysis.

Twig analysis (TA) - Measurements were made, on both branches, of yearly growth of the main leader or primary shoot between the girdle rings from 1996 to 1987 and of representative 5 and 10 year secondary shoots. Each subsidiary shoot of 1 to 10 years was recorded according to its age and hence year of bud formation, its subdivision (secondary, tertiary, quarternary etc) and whether it was acute, i.e. $\leq 40°$ to the parent stem or not. The pooled count gave the total number of subsidiary shoots per year and the percentage of acute shoots per tree. For statistical assessment the mean value of each parameter per tree was used.

Abiotic factors - Meteorological data was based on that of the Radcliffe station at Oxford. Soil moisture deficit was calculated by the Meteorological Office, using the MORECS system for deciduous trees and for soil of medium available water capacity. A highly significant correlation was found between the calculated values and the percentage soil moisture measured on site at Wytham Woods by the Environmental Change Network laboratory from July 1993 to December 1996. The cumulative total of the monthly maximum soil moisture deficit (mm) was used as a measure of the water stress over the given periods. Ozone data, monitored by AEA Technology about 20 km south of Wytham near Harwell, was provided as the accumulated ozone exposure above 40 ppb.h (AOT40). To calculate AOT40, the hourly concentrations above 40 ppb are summed for daylight hours over the period of concern (Fuhrer, Skarby & Ashmore 1997).

Statistical tests - The primary and secondary growth and per cent acute shoots (converted to arcsine) for the trees in the two compartments showed normal distributions. ANOVA tests showed highly significant ($P<0.001$) variation between trees and between years, apart from three exceptions with 25C data on the border of significance. Paired t tests for primary and secondary growth, and Chi-squared tests for the number of acute compared with non-acute shoots, were used to test the statistical significance of differences between specific years. Linear correlation coefficients were used to test the significance of the relationships between growth and abiotic factors (Minitab 12.2 computer programme). There was significant correlation between the abiotic factors and so it was inappropriate to use multiple regression analysis. An exception was for soil moisture deficit of the same year as the tree response (y) and for the previous year (y-1), since these showed no significant correlation

Results

Tree growth patterns

Primary growth. All nine trees of 55C showed a growth reduction from 1988 in the dry years of 1989 and 1990 (P<0.001; Figure 1). There was gradual recovery to the 1988 level by 1995. In marked contrast, the nine 25C trees showed little change until 1994 when the growth was significantly reduced from 1988 (P=0.006). Examination of the responses of individual trees suggested that this compartment should be separated into four trees (25CI) with some primary growth reduction in 1989 and 1990 compared to 1987/88, and five trees (25CII) with no reduction between these years, but some growth reduction in 1994-96 (Figure 1). This division was justified since there was no correlation between the primary growth of 25CI and 25CII (R^2=0.026).

Secondary growth. Secondary growth was more sensitive to environmental factors than primary. In the 55C trees, there was incomplete recovery, compared with 1987/88, by 1994, and some decline in growth in 1995, which had a high soil moisture deficit (Figure 1). 25CI trees also showed growth reduction in 1989 and in 1990 this was significant, compared with 1988. In contrast, 25CII trees showed gradual reduction of secondary growth, which became significantly different from 1988 in 1993, with a further significant reduction in growth in 1994.

Number of subsidiary shoots and percentage acute shoots. In line with the good primary and secondary shoot growth in 1988 the percentage of acute shoots was lowest in 1988 (Figure 2). It was found previously that low numbers of acute shoots were associated with young healthy trees (Stribley 1996). The control tree (RT=0) in each Wytham Wood compartment had low levels of acute shoots for all years, except with the 25C tree in 1990 where there was a significant increase compared with 1988 (P<0.01). Thus 1988 was used as the baseline year. All 55C trees showed a significant increase in the percentage of acute shoots in 1989 or 1990, compared with 1988; there was then a gradual return to a low level in 1994, but an increase occurred again in 1995 (Figure 2). Three trees failed ever to return to the baseline 1988 level, indicating a long term effect of the stress conditions of 1989/90. The 25CI trees had rather high 1988 per cent acute shoots and so increases would be less likely to be demonstrated; this was indeed the case (Figure 2). In the 25CII trees, the results were variable, showing some increases from 1988 in the early drought years and in 1995. The mean value in 1995 was significantly higher than that in 1988 and in 1994. As the trees grow, there are inevitably more young shoots than older ones. It was therefore interesting that, in both 25CI and 25CII, there was a reduction in the number of shoots from 1993 (Figure 2), which reflects the reduced primary and secondary growth from 1993 to 1994 described above.

Correlations between tree growth patterns and abiotic variables

Primary growth. Growth in compartment 55C showed a significant negative correlation with soil moisture deficit, with the highest correlation coefficient obtained using months January to September (Table 1). There was no significant effect of the previous year's soil moisture deficit in months January to September, but combining soil moisture deficit in y and y-1 in a multiple regression improved R^2 to 0.774 (P<0.001). In marked contrast, growth in compartment 25C overall showed no significant correlation with soil moisture deficit. 25CI trees did show a correlation with soil moisture deficit in months March to September which was just significant (P=0.05), but growth in 25CII showed no significant correlation with soil moisture deficit.

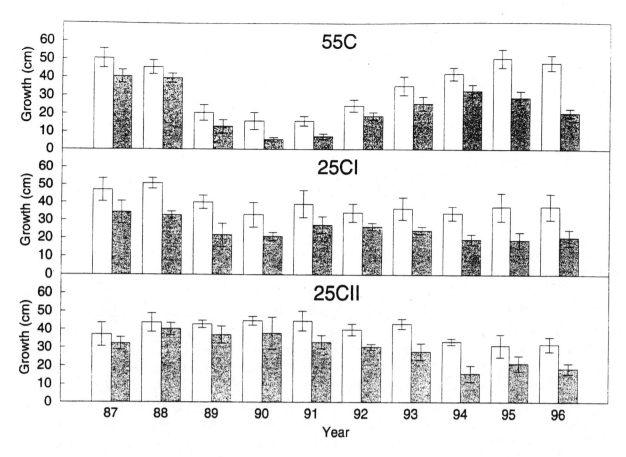

Figure 1. Mean yearly growth of primary and secondary (shaded column) shoots +/-SE in compartments 55C (n = 9) and 25C, sub-groups 25CI (n = 4) and 25CII (n = 5).

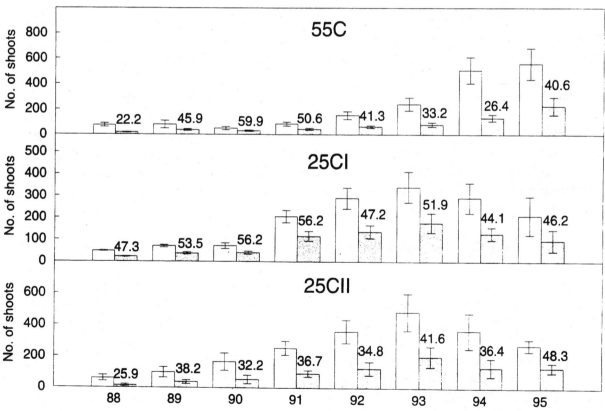

Figure 2. Mean number of subsidiary shoots and number acute (shaded column) +/-SE in compartments (see caption of Figure 1). The percentage of acute shoots are shown above the shaded columns.

Primary growth in compartment 55C showed no significant correlation with the ozone measure AOT40 (Table 1). In compartment 25C, however, there was a significant negative correlation with AOT40, calculated for both months March to September and June to September. This was due to the 25CI trees, as the 25CII trees showed no significant correlation. Thus 25CI trees had significant negative correlation with AOT40 in months June to September but there was higher correlation for March to September (R^2=0.798; P=0.001). Testing of residuals from the linear regression between 25CI primary growth and AOT40 in months March to September against soil moisture deficit over any period showed very low R^2 values; in contrast, testing of residuals from the regression between 25CI primary growth and soil moisture deficit in months January to September showed a significant correlation with AOT40 in months March to September. Thus, it appears that high ozone concentrations are associated with reduced primary growth in sub-group 25CI trees.

Secondary growth Trees in compartment 55C showed a significant correlation between secondary growth and soil moisture deficit in both months January to September and March to September (Table 1). Multiple regression with soil moisture deficit for months March to September in both y and y-1 increased R^2 to 0.799 (P<0.001). There was no significant correlation with soil moisture deficit for 25C overall, but secondary growth in 25CI showed a significant negative correlation with soil moisture deficit in months June to September (R^2=0.731; P=0.002). Compartment 55C had a significant negative correlation between secondary growth and AOT40 in both months June to September and March to September (Table 1). There was also a significant negative correlation with growth in 25C, for months June to September; as for primary growth, this was due to the response of the 25CI trees, which showed significant negative correlation with AOT40 in months March to September but June to September was highest (R^2=0.769; P=0.001). However, tests on the residuals both for 55C and 25C showed that, unlike for primary growth, no significant relationships with soil moisture deficit and AOT40; hence, the correlations of soil moisture deficit and AOT40 could not be separated.

Percentage of acute shoots The percentage of acute shoots in 55C showed the highest significant positive correlation with soil moisture deficit in months March to September (R^2=0.749; P=0.005), and multiple regression of soil moisture deficit in y and y-1 significantly improved the fit (R^2=0.914, P<0.001). However, no significant correlation with soil moisture deficit was found for 25C trees, or for the two sub-groups (Table 1). There was no significant correlation between AOT40 and the percentage of acute shoots in 55C but for 25CII months June to September correlation was nearly significant (P=0.057).

Association of Roloff twig canopy score with percentage of acute shoots

The mean percentage of acute shoots for trees with RT canopy scores of 0 (including a control tree) or the border of 0-1 were compared with the mean for trees with a RT score of 1 or 1→0, for both compartment 55C and 25C. The years chosen were 1988, to represent the background level, and 1994, which was the year of best recovery for most trees. For comparison with earlier studies (Stribley 1996), the Wytham Woods tree TA data was put in the 'basic' TA format by counting all two year shoots produced on the primary shoot for the five years 1990-94. Whichever year was tested by Chi-squared, the number of acute shoots against non-acute were significantly different between RT=0/0-->1 and RT=1/1-->0 (P values <0.05 to <0.001).

Table 1. Linear correlation between tree responses and abiotic criteria

Tree response*	months	SMD Rsq	P value	months	AOT40 Rsq	P value
55CM1	1-9	0.608	0.008			
"	3-5	0.517	0.019	3-5	0.101	
"	6-9	0.043		6-9	0.147	
"	3-9	0.402	0.049	3-9	0.205	
25CM1	3-5	0.049		3-5	0.144	
"	6-9	0.320	0.088	6-9	0.424	0.041
"	3-9	0.111		3-9	0.474	0.028
25CIM1	3-5	0.040		3-5	0.379	0.058
"	6-9	0.361	0.066	6-9	0.585	0.010
"	3-9	0.398	0.050	3-9	0.798	0.001
25CIIM1	3-9	0.004		3-9	0.047	
55CM2	1-9	0.593	0.009			
"	3-5	0.295		3-5	0.087	
"	6-9	0.267		6-9	0.421	0.043
"	3-9	0.609	0.008	3-9	0.416	0.044
25CM2	3-5	0.191		3-5	0.011	
"	6-9	0.301	0.101	6-9	0.450	0.034
"	3-9	0.033		3-9	0.323	0.087
25CIM2	3-5	0.033		3-5	0.051	
"	6-9	0.731	0.002	6-9	0.769	0.001
"	3-9	0.363	0.065	3-9	0.620	0.007
25CIIM2	6-9	0.092		6-9	0.211	
"	3-9	0.004		3-9	0.134	
55CMA	1-9	0.532	0.040			
"	3-5	0.385	0.101	3-5	0.181	
"	6-9	0.341	0.128	6-9	0.286	
"	3-9	0.749	0.005	3-9	0.380	0.101
25CMA	3-5	0.002		3-5	0.000	
"	6-9	0.402	0.092	6-9	0.426	0.079
"	3-9	0.312	0.150	3-9	0.210	
25CIMA	3-5	0.129		3-5	0.010	
"	6-9	0.070		6-9	0.017	
"	3-9	0.192		3-9	0.002	
25CIIMA	3-5	0.019		3-5	0.009	
"	6-9	0.328	0.138	6-9	0.478	0.057
"	3-9	0.161		3-9	0.296	

*M1, M2, MA - Means primary and secondary growth and per cent acute shoots from each compartment 55C and 25C (n=9). 25CI(n=4) and 25CII(n=5) sub groups of 25C. SMD - summed total of monthly maximum soil moisture deficit (mm); AOT40 - accumulated ozone total >40 ppb.h. All primary and secondary correlations were negative and acute shoots were positive.

Discussion

Quantitative twig analysis (TA) demonstrated primary and secondary shoot growth and the total number of subsidiary shoots were suppressed in drought years, while the percentage of shoots growing at $\leq 40°$ increased. In the 55C compartment, all these responses were significantly correlated with soil moisture deficit. Evidence from open-top chamber experiments has demonstrated that ozone exposure reduced growth of well watered, but not droughted, beech seedlings (Pearson & Mansfield 1994). The present primary growth results appear to be consistent with these experimental findings, in that no ozone correlation was found in 55C, where soil moisture deficit had significant correlation, but was found in 25C, where soil moisture deficit had much less correlation. Possibly the shallow soil and wide spacing of the 55C trees made them more susceptible to drought than those in 25C which were close together and where the soil was deeper.

A series of workshops for the UNECE in Europe, discussed in Fuhrer, Skarby & Ashmore (1997), have proposed a critical AOT40 level of 10,000 +/- 5,000 ppb.h. This was based on a 10 per cent biomass reduction of seedlings of beech (which was the most sensitive species) in chamber experiments. AOT40 calculated over a six months growing period showed values of around 10,000 ppb.h over the period 1989-1995 at Harwell (data not shown). Our data suggest that this value of the critical level is too high to prevent significant effects on the primary growth of mature woodland beech but more research is needed to confirm the applicability of this critical level to mature trees. Recently, Braun, Rihm and Fluckiger (1999) reported from 1992-95 ozone was negatively correlated with the stem increment of mature beech in Switzerland and proposed a reduction of the present AOT40 critical level.

The close correlation of percentage of acute shoots in compartment 55C with soil moisture deficit emphasises changing environmental factors as strongly influencing acute shoot formation rather than genetics or light exposure. Examination of the beech saplings in the Forestry Commission open-topped chamber experiment (Durrant *et al.* 1992) where ozone pollution was associated with significantly reduced growth, showed increased percentage of acute shoots in the outside group (P<0.02) compared to the chamber groups; there was 21 per cent greater number of acute shoots in the ambient air group compared to the filtered air group although there was variation between the saplings (Stribley unpublished observations). These results and the 25CII correlation indicate that ozone could be affecting acute shoot formation but further research is needed to clarify this.

In previous TA of beech trees of differing ages and sites, there was 88% agreement of the TA-based categorisation and the canopy RT score assessed before the twigs were obtained. For trees <70 years old, the percentage of acute shoots was a distinguishing criterion between normal twig pattern (RT=0) and trees showing early decline, with a RT score of 1 corresponding to ≥21 per cent acute shoots. Wytham Wood trees were selected to cover early changes and so most were on the border between RT score 0 and 1. A decision as to whether the appearance was RT= 0-1 or RT=1-0, was difficult, because of variation between branches leading to potential sampling error. Furthermore, the close proximity of 25C trees meant assessment of RT was through lower twigs and so underscored. With the more widely spaced trees of 55C, a good view of the top canopy was possible and all 10 trees RT scores correlated with the 1988 percentage of acute shoots within the predicted levels described above.   In spite of the difficulties in RT assessment the increase in acute shoot formation of trees with raised RT scores was significantly different from low RT scores whether 1988, 1994 or the basic method was used.

Thus, overall the percentage of acute shoots reflected the RT canopy score. The present results provide validation for the use of the RT winter assessment to monitor changes in the health of trees in the countryside. Long term monitoring demonstrated that young trees are deteriorating (Stribley 1999), with RT scores being found which would not be expected until around 140 years old (Roloff 1985). This is of particular concern and emphasises the importance both of continued monitoring and further research to understand the causes of the recorded changes in canopy growth patterns. The TA results support the interpretation that both drought and ozone pollution may be important causative factors in the long term decline of young woodland trees observed using Roloff winter twig pattern assessment. The observation that acute shoots are formed in response to the stress of drought years should provide a valuable tool to examine detailed tree responses to a changing climate and pollution environment.

Acknowledgements

The work was in part funded by a British Ecological Society grant.

References

BRAUN, S., RIHM, B. & FLUCKIGER, W. 1999. Growth of mature beech in relation to ozone: an epidemiological approach. *In: Critical Levels for Ozone - Level II* . UNECE Workshop April 1999 Report. Switzerland, Swiss Federal Research Station for Agroecology and Agriculture, pp 77-80.

DAVIDSON, S.R., ASHMORE, M.R. & GARRETTY, C. 1992. Effects of ozone and water deficit on the growth and physiology of *Fagus sylvatica. Forest Ecology & Management*, **51**, 187-93.

DURRANT, D.W.H., WADDELL, D.A., BENHAM, S.E. & HOUSTON, T.J. 1992. *Air quality and tree growth in open-top chambers.* Forestry Commission Information Note No. 208. Edinburgh, Forestry Commission.

FUHRER, J., SKARBY, L. & ASHMORE, M.R. 1997. Critical levels for ozone effects on vegetation in Europe. *Environmental Pollution*, **97**, 91-106.

LONSDALE, D., HICKMAN, I.T., MOBBS, I.D. & MATTHEWS, R.W. 1989. A quantitative analysis of beech health and pollution across southern Britain. *Naturwissenschaften*, **76**, 571-573.

PEARSON, M. & MANSFIELD, T.A. 1994. Effects of exposure to ozone and water stress on the following season's growth of beech (*Fagus sylvatica* L.). *New Phytologist*, **126,** 511-515.

POWER, S.A. 1994. Temporal trends in twig growth of *Fagus sylvatica* L. and their relationships with environmental factors. *Forestry*, **67**, 13-30.

POWER, S.A., ASHMORE, M.R. & LING, K.A. 1995. Recent trends in beech tree health in southern Britain and the influence of soil type. *Water, Air and Soil Pollution*, **85**, 1293-1298.

ROLOFF, A. 1985. Schadstufen bei der Buche. *Der Forst und Holzwirt*, **40**, 131-134.

STRIBLEY, G.H. 1993. Studies on the health of beech trees in Surrey, England: relationship between winter canopy assessment by Roloff's method and twig analysis. *Forestry*, **66**,1-26.

STRIBLEY, G.H. 1996. Quantitative twig analysis of beech (*Fagus sylvatica* L.) trees of varying ages and health. *Forestry*, **69**, 271-273.

STRIBLEY, G.H. 2000. Long term monitoring of the health of beech (*Fagus sylvatica* L.) trees of all ages in Surrey. *In:* KIRBY, K.J. & MORECROFT, M.D., eds. *Long-term studies in British woodland.* pp 110-111. (English Nature Science No. 34.) Peterborough, English Nature.

LINKING CLIMATE AND BIOLOGICAL MONITORING: THE EFFECTS OF DROUGHT AT WYTHAM WOODS

M.D. Morecroft
Institute of Terrestrial Ecology, Oxford University Field Laboratory, Wytham, Oxford OX2 8QJ
e.mail: mdm@ceh.ac.uk

Summary

The summer of 1995 was the second driest and third warmest on record in the Oxford area. Data from the Environmental Change Network (ECN) monitoring programme at Wytham Woods were examined to detect impacts of the drought on vegetation, butterflies and ground beetles. Despite trees losing their leaves early and ground vegetation dying back, no major change in vegetation composition or tree mortality was found. In contrast grassland monitoring plots saw an increase in species number as ruderal species colonised gaps which opened up in the sward. Some animal populations increased in 1995, others decreased. In general terms most beetle species declined whilst most butterflies increased. There were however exceptions. In particular, speckled wood butterflies (*Parage aegeria*) decreased and the beetle *Nebria brevicollis* increased. The effects of this drought disappeared within four years, but a series of similar summers would give cause for concern. Current predictions of climate change suggest that such a sequence of events is becoming more likely.

Introduction

Monitoring of the natural environment has often been a matter of isolated studies of single aspects of the environment - one person recording butterflies here, another measuring tree growth there and some else making measurements of climate 20 miles down the road. Some of these studies have been very valuable, but the linkages between different aspects of the environment are important if we are going to fully understand ecological systems. Understanding these linkages is much easier if the monitoring of different environmental variables is integrated together at the same sites. This paper describes one approach to integrated monitoring, that of the UK Environmental Change Network (ECN).

The ECN was set up in 1992 by a consortium of sponsoring organisations (see acknowledgements). The measurements made within ECN are diverse and span both the physical and biological aspects of the environment (Table 1). The main theme is to link ecological variables with those factors that are likely to be driving change in ecosystems, especially climate, air pollution and land management. The Network is really 2 networks, a freshwater one composed of lake and river sites and a terrestrial one. There are presently 12 terrestrial sites in ECN; they are located throughout Great Britain and Northern Ireland and cover a wide variety of vegetation types and land uses. Not all have woodland, but the sites at North Wyke, Porton, Wytham, Alice Holt, Rothamsted and Hillsborough do. At Wytham and Alice Holt the woodland is the main land cover on the site. Further details of ECN can be found on the ECN web site, http://www.nmw.ac.uk/ecn.

Table 1. Summary of ECN core measurements

Meteorology

Atmospheric Chemistry (nitrogen dioxide + ammonia)

Surface water flow & chemistry (concentration of major ions).

Soil solution chemistry (concentrations of major ions).

Precipitation chemistry (concentrations of major ions).

Soil characteristics (baseline survey, 5 year and 20 year recording in permanent plots).

Vegetation composition (baseline survey, 3 year and 9 year recording in permanent plots).

Vertebrate populations (birds, rabbits, deer, bats, frogs).

Invertebrate populations (butterflies, moths, ground beetles, spittlebugs, cranefly larvae).

Site Management

This paper focuses on Wytham Woods and looks at how the impacts of an extreme climatic event - the drought of 1995 showed up across different aspects of the woodland ecosystem. The summer (June - August) of 1995 was record breaking. It was the driest in the 229 year record for England and Wales and August 1995 was second warmest month ever in the Central England Temperature record, which goes back to 1659 (Marsh 1996). A long term monitoring study is essential to detecting such impacts as we can look at data before, during and after the event and ideally set it in a context of similar events in the past.

Site Description

Wytham Woods have been owned by Oxford University since 1943 and has been the site of many pioneering studies, particularly in animal ecology. They cover an area of approximately 400 ha and include several different woodland types. Approximately one third of the site is ancient woodland, which was managed as coppice with standards for many hundreds of years and then abandoned over the course of the twentieth century. The coppice stools are mostly hazel and the standards mostly oak. Another third of the site was formerly wood pasture, pasture or arable land, but has naturally regenerated to woodland at various different times from the eighteenth century up to quite recent times. Ash tends to dominate the older areas of this secondary woodland, sycamore the more recent. All of these areas are subject to minimum intervention at present and are W8, *Fraxinus - Acer campestre - Mercurialis* woodland in the National Vegetation Classification (NVC) (Rodwell 1991); the ancient areas tending to be W8a the *Primula vulgaris - Glechoma hederacea* community and the newer areas W8e, the *Geranium robertianum* sub- community. There are also plantations of various species, including beech, oak, larch and Norway spruce. Of these, beech is the most significant and produces W12 and W14 communities. The vegetation and stand structure and changes between 1974 and 1991 years have been described by Kirby, Thomas and Dawkins (1996) and Kirby and Thomas (1999). There are also small areas of semi-natural grassland and scrub.

The grasslands include an ex-arable calcareous grassland (MG1d, *Arrhenatherum elatius* grassland, *Pastinaca sativa* sub-community), old semi-natural mesotrophic grassland (*MG5b Cynosurus cristatus - Centaurea nigra* grassland, *Galium verum* sub-community) (Rodwell 1992) and mosaics of old calcareous grassland with scrub. Data on grassland in this paper also include one plot in intensively managed agricultural grasslands (MG7 *Lolium perenne* grasslands) on adjoining farmland, which is also owned by Oxford University.

Methods

Climate

ECN monitoring methodology has been described in detail by Sykes and Lane (1996). A brief summary of the methods relevant to this paper is given here, together with site specific details and additional work carried out in 1995 itself.

Climate was monitored with an automatic weather station, which stores data on an hourly basis. These automatic measurements started in 1992, but we are fortunate in that a very long term series of climate data have been collected in Oxford itself at the Radcliffe Observatory, just 5 km from the site. Precipitation records go back to 1767, temperature to 1815.

Trees

42 Permanent plots, 10 m. x 10 m. square, were established to record tree growth in 1993, using standard ECN methodology. These were selected randomly but based on a grid system which was already present (Dawkins & Field 1978). Within each plot up to 10 trees were marked and diameter at breast height (DBH) and height recorded. DBH was intended to be recorded every 3 years so these quadrats were revisited in 1996 one year after the drought and again in 1999 (an additional survey was carried out on most plots in 1998 also). A total of 288 trees were marked and recorded.

One of the most obvious impacts of the drought was that many deciduous species lost their leaves early in 1995. To quantify which species were most affected, a series of transects (total length approximately 7.5 km) were walked in early September, counting how many individuals of each species, had suffered total leaf loss (defined as less than an estimated 1% leaves remaining). At an early stage it became apparent that elder (*Sambucus nigra*) was particularly affected, but the defoliation was somewhat patchy. For just elder, the number of bushes which had not suffered total leaf loss was also counted along the transects and the proportion of defoliated trees compared with soil type (surveyed for ECN in 1992; Beard 1993).

Ground vegetation

Ground vegetation was recorded, in 10 plots, representing the different main NVC communities at Wytham, woodland and grassland, in 1994, 1996, 1997, 1998 and 1999. The ECN 'fine grain' method was used, this is based around 10m x 10m plots containing 10 randomly distributed, permanent quadrats of 0.4m x 0.4m. Recording was intended to be repeated every 3 years, but additional recording was carried out as part of a project to study annual variability in vegetation.

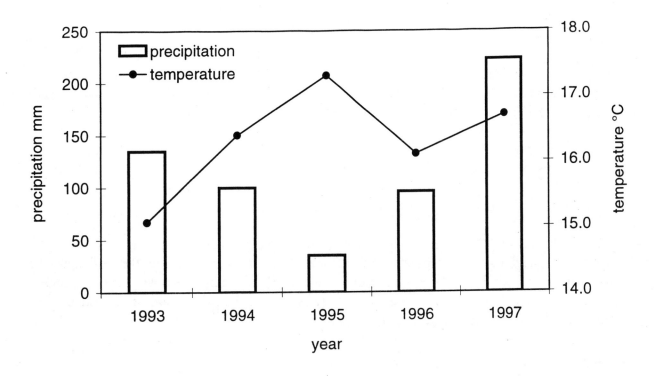

Figure 1. Mean temperature and total precipitation during summer months (June, July, August) at Wytham 1993-97

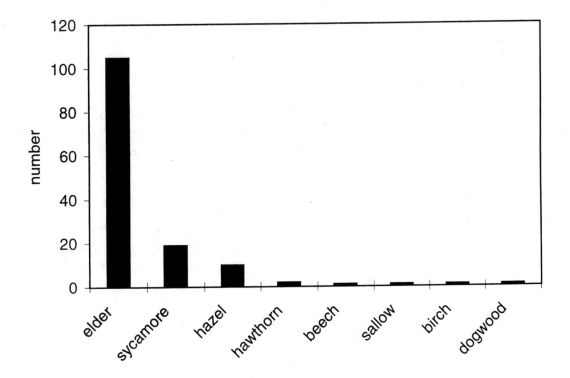

Figure 2. Premature leaf loss in tree species at Wytham. Number of individuals of each species found on the transect, recorded as showing at lest 99% leaf loss

Invertebrates

A variety of invertebrate groups were sampled (Table 1); only two are presented here, butterflies and ground beetles. Both are subject to a detailed monitoring methodology, which allows comparisons to be made between contrasting species.

Butterflies were recorded each week, from the beginning of April to the end of September, along a 3.5 km transect, using the method of the ITE Butterfly Monitoring Scheme (BMS, Pollard & Yates, 1993), which ECN has also adopted. The transect runs along the northern boundary of Wytham Woods, beside the River Thames and through mixed agricultural land.

Ground beetles (Carabidae) were sampled with pitfall traps, arranged in 3 transects of 10 traps in contrasting habitats, with traps spaced 10 m apart. The traps are run from the beginning of May to the end of October and are emptied every two weeks. Data from two woodland transects are presented here, one in an area of ancient semi-natural woodland, one in an old beech plantation (approximately 200 years old).

Results

Climate

Figure 1 shows temperature and rainfall in the summer months at Wytham between 1993 and 1997. 1995 clearly stands out, being both warmer and drier than other years. Records from the Radcliffe meteorological station showed the summer of 1995 to be the third warmest since 1815 and the second driest since 1767. It is reasonable to assume this is the same for Wytham: comparison of climate data from Wytham and Oxford since 1993, showed that the two data sets were very similar. The main difference is that Oxford is slightly warmer in winter, because of its urban location and 100m lower altitude. Localised summer thunderstorms can also produce surprisingly different precipitation figures for Wytham and Oxford, but this did not happen in 1995. There have been large variations in precipitation in the 1990's. The winter of 1994 - 95 was extremely wet (272 mm, for December, January and February combined and the sixth wettest on record in Oxford) which meant that soils were unusually wet when the summer of 1995 began. The soil dried out over the course of the summer and remained relatively dry through 1996 and the first part of 1997 as below average precipitation continued through this period.

Trees and shrubs

The results for leaf loss are shown in Figure 2. The biggest impact was found to be on understorey shrubs and young trees, particularly elder, as anticipated. The proportion of elder trees showing total leaf loss ranged from 19% on thin, stony soils (Moreton and Sherborne Series in the Soil Survey classification for England and Wales) on the higher ground to 58% on deep clay soils (Denchworth and Evesham series). The greater leaf loss on the deeper soils was surprising but very significant (P < 0.001; Pearson Chi-square test, Systat 8.0), it may reflect the tendency for the deep clay soils to shrink and crack severely in dry conditions.

Despite the premature leaf fall in 1995, relatively few trees and shrubs died between 1993 and 1996. All of the monitored trees in the canopy survived; in the understorey, 9 individuals died, of which 4 were elder. In the subsequent 3 years (1996 - 1999) 7 further trees died, of which one was elder.

Ground vegetation

During the summer 1995 itself, substantial die back of the ground vegetation was observed, for example by the 31 July most of the dog's mercury *Mercurialis perrenis* had wilted. However, no change was observed in the species recorded before and after the drought in the permanent plots. This contrasts with the situation in grassland plots, where the number of species increased after the drought (Figure 3), as a consequence of ruderal forb species colonising gaps which opened up as grasses died back. This was particularly a feature of the ex-arable and agricultural grasslands.

Invertebrates

Most butterfly species had high populations in the summer of 1995. This could be clearly seen from examining the whole record from 1993 - 1998, but in order to quantify this effect, 1995 numbers were compared with those in 1994 (very similar results are obtained by using mean numbers for the whole record as a reference rather than 1994 data). More species increased between 1994 and 1995 than decreased or did not change (Table 2); the meadow brown *Maniola jurtina*, the most common butterfly on the transect, is a good example (Figure 4). There was also no crash in numbers in 1996, with most species still more abundant than in 1994. One notable exception is the distinctive woodland species, the speckled wood, which decreased in 1995, with low numbers continuing into 1996 (Figure 4).

The ground beetles contrast with the butterflies. In this case more species decreased from 1994 to 1995 than increased (Table 2), and the pattern also continued into 1996. The difference between butterflies and ground beetles was significant in the 1994 - 1995 comparison ($P < 0.001$; Pearson Chi-square test, Systat 8.0) but not for 1994 - 1996. There were however some beetle species which increased in number, including the most numerous species recorded by ECN at Wytham, *Nebria brevicollis*.

Table 2. Number of species increasing and decreasing over the periods 1994 - 5 and 1994-6 for butterflies and carabid beetles

	Increase	Decrease	No change
Butterflies 1994 - 1995	14	5	2
Ground beetles 1994 - 1995	3	13	8
Butterflies 1994 - 1996	11	7	3
Ground beetles 1994	8	11	5

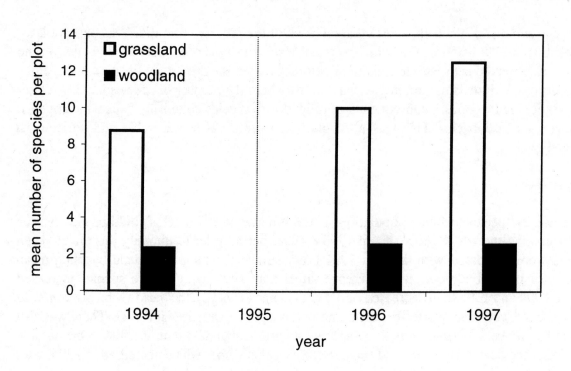

Figure 3. Mean number of species of forb in woodland and grassland plots 1994 - 1998.

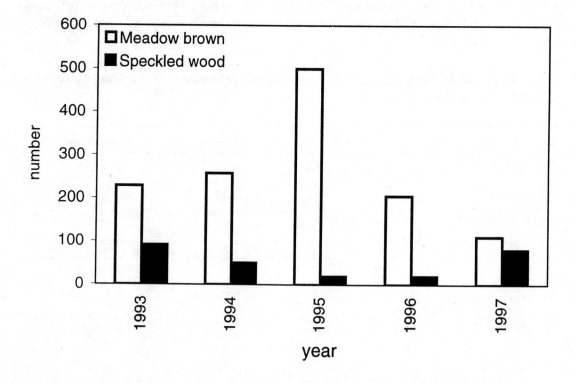

Figure 4. Numbers of meadow brown and speckled wood butterflies recorded on the Wytham butterfly transect, 1993 - 1998.

Discussion

The drought of the summer of 1976 is often regarded as the most severe in recent times and a number of studies revealed effects which can be compared with the impacts of the 1995 drought reported here. Long term monitoring of trees at Lady Park Wood on the Wales - England border showed that a large number of old beech and young birch trees died following the 1976 drought (Peterken & Jones 1989; Peterken & Mountford 1996), changing the character of parts of the wood. No comparable effect was seen at Wytham in 1995 or the years following; in fact no death of marked canopy trees has been seen since the start of the ECN programme. There was death in the understorey, but it is hard to attribute this to the drought with confidence as higher mortality would be expected in young trees and we do not have comparable data from the pre-drought period. It is however interesting that elder was the tree showing both the highest leaf loss in 1995 and the highest mortality between 1993 and 1996. One crucial difference between the summers of 1976 and 1995 was that the 1976 summer was preceded by a period of very dry conditions, whereas the winter of 1994/95 was unusually wet, which ensured soil water content remained reasonably high at deeper levels.

The increase in ruderal species in grassland has parallels with results from grassland verges at Bibury, Gloucestershire, which have been monitored since 1958. Dunnett *et al.* (1998) reported a dramatic increase in the abundance of cow parsley (*Anthriscus sylvestris*), a competitive-ruderal species, in 1977. In both cases the creation of gaps following the die-back of dominant grass species allowed the opportunity for this to take place. A drought simulation experiment carried out at Wytham has shown the same processes occurring (Clarke, Masters & Brown, in press). Because canopy trees generally survived the 1995 drought, there was no equivalent gap creation in woodland and hence no opportunity for ruderal species to increase. This may be an intrinsic difference between woodland and grassland, reflecting the higher temperatures in grass swards than woodland canopies (e.g. Morecroft, Taylor & Oliver 1998). It may also be explained in terms of a higher proportion of stress tolerant species (Grime 1976) in the woodland: interestingly, old grasslands with a higher proportion of 'stress tolerators', appeared to be less affected than more recent grasslands.

The Butterfly Monitoring Scheme only started in 1976, but an analysis for the period 1976 - 1986 showed that most species tended to increase during hot, dry, summers although a few species, including the speckled wood decreased (Pollard 1988). As at Wytham, national BMS data showed that most species increased from 1994 to 1995, with the few decreasing ones, again including the speckled wood (Pollard & Greatorex-Davies 1995). Most butterflies tend to be found in open areas and are adapted to relatively high temperatures and dry conditions; the adults in particular need dry, warm conditions to fly. Butterflies can, as a group, be regarded as pre-adapted to hot, dry summers (in a British context); the speckled wood is an exception, being one of relatively few species adapted to the damp shady conditions under woodland canopies. In contrast most ground beetles and their larvae are found in the relatively cool, damp conditions of the litter layer on the soil surface and are not adapted to surviving drought. This is especially true in woodlands where warm, dry conditions in the litter layer are extremely rare.

The results presented here are only a small subset of all ECN data. Another interesting aspect is the impact on stream nitrate concentrations (Morecroft *et al.* in press). Nitrate concentrations in streams draining both woodland and agricultural land at Wytham showed an increase following the summer of 1995, which persisted until the winter of 1997/98 when soil water contents finally recovered to pre-drought levels. This is partly because of a concentrating effect of reduced water flow in streams and partly because of enhanced mineralisation and nitrification rates in the soil, which may, in turn result from changes in the chemical composition of litter.

Overall the summer of 1995 did not substantially alter the character of Wytham Woods and woodland vegetation appears to have been particularly resilient. There were important effects on some particular species and ecosystem processes, but recovery seems to be complete as I write in 1999. Current predictions suggest that climate change will lead to increased temperatures and drier summers in the South East of Britain (Hulme & Jenkins 1998). The big question, is what happens if we have a series of dry summers? If we reach the point at which canopy trees start to die, much larger changes than those recorded here could ensue. After seven years, ECN is relatively young for a long-term monitoring scheme. This has however proved long enough to characterise the impacts of the 1995 summer drought with reference to pre-drought conditions and to assess recovery in the following years. This work has however been made much more meaningful by the availability of longer term data, from other sources. This is especially true of meteorological recording, without which it would not have been possible to recognise the degree to which 1995 was an exceptional year in the first place. The long tradition of monitoring the weather in a scientific way, should be an inspiration for ecologists and foresters as we endeavour to establish and maintain long term studies. In a changing climate these studies will be increasingly important if management decisions are to be made on a sound basis.

Acknowledgements

ECN research at Wytham is funded by the Natural Environment Research Council. The other sponsors of ECN are: Biotechnology & Biological Sciences Research Council, Countryside Council for Wales, The National Assembly for Wales, Defence Evaluation and Research Agency/ Ministry of Defence, Department of Agriculture for Northern Ireland, Department of the Environment for Northern Ireland, Department of the Environment, Transport and the Regions, English Nature, Environment Agency, Forestry Commission, Ministry of Agriculture, Fisheries & Food, Scottish Environment Protection Agency, Scottish Executive Rural Affairs Department, Scottish Natural Heritage.

I am also grateful to the School of Geography, Oxford University for providing data from the Radcliffe Meteorological station and all those who have played a part in collecting the data that are presented here, especially Michèle Taylor.

References

BEARD, G.R. 1993. *The soils of Oxford University Field Station, Wytham.* Silsoe, Soil Survey and Land Research Centre.

DUNNETT, N.P., WILLIS, A.J., HUNT, R. & GRIME, J.P. 1998. A 38 – year study of relations between weather and vegetation dynamics in road verges near Bibury, Gloucestershire. *Journal of Ecology,* **86**, 610 – 623.

CLARKE, I.P., BROWN, V.K. & MASTERS G.J. (In press) Calcareous grassland dynamics in a manipulative climate change experiment. *Oikos.*

DAWKINS, H.C. & FIELD, D.R.B. 1978. *A Long-term surveillance system for British woodland vegetation.* (Occasional Papers of the Oxford Forestry Institute No. 1). Oxford, Oxford Forestry Institute.

GRIME, J.P. 1979. *Plant strategies and vegetation processes.* New York, John Wiley.

HULME, M. & JENKINS, G.J. 1998. *Climate change scenarios for the UK: scientific report.* UKCIP technical report no. 1. Norwich, Climatic Research Unit.

KIRBY, K.J., THOMAS, R.C. 1999. Changes in the ground flora in Wytham Woods, southern England, 1974-1991, and their implications for nature conservation. Peterborough, *English Nature Research Reports*, No. 320.

KIRBY, K.J., THOMAS, R.C. & DAWKINS, H.C. 1996. Monitoring of changes in the tree and shrub layers in Wytham Woods (Oxfordshire), 1974 - 1991. *Forestry*, **69**, 19-334.

MARSH T.J. 1996. The 1995 UK drought – a signal of climatic instability? *Proceedings of the Institution of Civil Engineers: Water, Maritime & Energy*, **118,** 189 – 195.

MORECROFT, M.D., TAYLOR, M.E. & OLIVER, H.R. 1998. Air and soil microclimates of deciduous woodland compared to an open site. *Agricultural and Forest Meteorology*, **90**, 141-156.

MORECROFT, M.D., BURT, T.P., TAYLOR, M.E. & ROWLAND, A.P. (In press). Effects of the 1995-1997 drought on nitrate leaching in lowland England. *Soil Use & Management.*

PETERKEN, G.F. & JONES, E.W. 1989. Forty years of change in Lady Park Wood: the young - growth stands. *Journal of Ecology,* **77**, 401 - 429.

PETERKEN, G.F. & MOUNTFORD, E.P. 1996. Effects of drought on beech in Lady Park Wood, an unmanaged mixed deciduous woodland. *Forestry*, **69**, 125- 136.

POLLARD, E. 1988. Temperature, rainfall and butterfly numbers. *Journal of Applied Ecology,* **25**, 819 - 828.

POLLARD, E. & YATES, T.J. 1993. *Monitoring butterflies for ecology and conservation.* London, Chapman & Hall.

POLLARD, E. & GREATOREX-DAVIES, J.N. 1995. *The Butterfly Monitoring Scheme. Report to Recorders.* Huntingdon, Institute of Terrestrial Ecology.

RODWELL, J.S. ed. 1991. *British Plant Communities Volume 1. Woodlands and Scrub.* Cambridge, Cambridge University Press.

RODWELL, J.S. (ed.) 1992. *British Plant Communities Volume 3. Grasslands and montane communities.* Cambridge, Cambridge University Press.

SYKES, J.M. & LANE A.M.J. 1996. *The United Kingdom Environmental Change Network: Protocols for standard measurements at terrestrial sites.* London, The Stationery Office.